THE EXISTENCE PRINCIPLE

PHILOSOPHICAL STUDIES SERIES

VOLUME 75

Founded by Wilfrid S. Sellars and Keith Lehrer

The titles published in this series are listed at the end of this volume.

THE EXISTENCE PRINCIPLE

QUENTIN GIBSON

Australian National University,
Canberra, Australia

KLUWER ACADEMIC PUBLISHERS

DORDRECHT / BOSTON / LONDON

A C.I.P. Catalogue record for this book is available from the Library of Congress.

ISBN 0-7923-5188-6

Published by Kluwer Academic Publishers,
P.O. Box 17, 3300 AA Dordrecht, The Netherlands.

Sold and distributed in North, Central and South America
by Kluwer Academic Publishers,
101 Philip Drive, Norwell, MA 02061, U.S.A.

In all other countries, sold and distributed
by Kluwer Academic Publishers,
P.O. Box 322, 3300 AH Dordrecht, The Netherlands.

Printed on acid-free paper

Printed in the Netherlands

CONTENTS

ACKNOWLEDGEMENTS

Papers which have developed into chapters of this book have been read over a number of years to seminars of the Departments of Philosophy at the Australian National University and at conferences of the Australasian Association of Philosophy. I wish to acknowledge the help towards a final version given by discussions in these sessions as well as individually by colleagues in both departments. I wish also to thank for their comments J.J.C. Smart and David Armstrong, both of whom have read sections of the manuscript, and Joyce Gibson for her invaluable critical attention to the content and the style of the whole.

Finally, I wish to acknowledge the arduous task performed by Reta Gear, of the Arts Philosophy Department's administration in processing hand-written manuscript into finished and duplicable typescript.

CHAPTER ONE

THE CONCEPT OF EXISTENCE

The world is full of many things, and there are those of us within it who are concerned to enquire about their nature and their relations to one another. In doing this we have already taken it for granted that they exist, for this after all is what we mean when we say that there is a world and that it is full of them.

If we are asked to explain in turn what we mean by saying they exist we can point to certain requirements to which such talk of their existence commits us. But there is one thing we cannot do and that is to state in any other terms what it is for them to exist. The concept of existence, in other words, is an absolutely elementary concept. The attempt to define it further involves one inevitably in circularity. The circularity is of course due to the fact that anything whatsoever in terms of which one might attempt to define it must itself exist. To say, for example, that to exist is to be perceived is to say that it is to exist as an object of perception. To say that to exist is to be spatio-temporal is to say that it is to exist in space and time. Hence despite appearances nothing is being said here about what it is for things to exist. Instead, theories are being produced about what kinds of thing exist, each claiming that they are all of a certain kind. Anyone who enquires about the nature of things may have to consider such theories, which we may call metaphysical theories. But they do nothing to illuminate further the concept of existence. If ontology is an enquiry which is supposed to do this, there is no such thing as ontology.

USES OF THE WORD 'EXIST'

These statements, it should be made clear, are not intended as statements about the way the word 'exist' is used, or should be used. No one doubts that it is used in different ways in different contexts, and it is not our business to say which is proper. In all but one of these uses, what I have said about the elementariness of the concept does not apply to it. That one use is a common and important one, and one which I have taken for granted in making my pronouncements. It is important to distinguish it from others, so that these others can be put on one side.

One of those others is that in which existence is opposed to essence, existence being attributed to the concrete particular as distinct from the abstract or the universal. Those who call themselves existentialists appear to rely on this usage, even making it more specific by attributing existence only to those concrete particulars which are independent of external influences. Then again existence is sometimes limited to the present time, as when it is said of past things which have decayed or been destroyed that they have ceased to exist. And as a final illustration we should mention a sense of 'exist' in which to exist is to be an item in some universe of discourse like the natural numbers or the characters in Shakespeare's plays. Among the natural numbers there exists in this sense a prime odd number greater than 3, but not an even one; and in the Shakespearean corpus Hamlet exists, but Faust does not.[1]

This variety of uses of 'exist' is matched by the variety of the uses of other words which are closely related to it such as 'be', 'be actual' and 'be real'. Often these are treated as synonymous with 'exist' whatever the meaning given to 'exist', though sometimes the uses drift apart, as when real coins and real friends are distinguished from the counterfeit and the deceitful.[2] But whether these uses diverge or not, there remains in each case one elementary use, and in that use 'reality', 'actuality' and 'being' are undoubtedly synonymous with 'existence' in its elementary sense.

In the case of all these words, fortunately, it is relatively easy to identify the use with which we are concerned. This is because of its very special character. It is the one use of the word, that is to say, which it is impossible to give an account of in any other terms without circularity. It is to be noted that the other uses which I have mentioned are ones of which I have in fact given such an account. I have pointed to certain classes of things with which what exists, in these senses, can be identified—the particular and concrete, the temporally present, the items in some specified universe of discourse. And in doing this I have been under no pressure to include the word 'exist', in the sense in question, in the account I have given.

It is true that the word 'exist' may occur in such an account. Thus in identifying existing with being particular and concrete, one might express oneself by saying that to exist is to exist as a concrete particular. Similarly, instead of saying that to exist is to be present, one might say that it is to exist in the present. But there would be no circularity here, since the second use of 'exist' in these statements is not the same as the first. In fact it is nothing other than our elementary sense, and our ability to include it merely shows that we are assuming all along that the particular and the present exist not only in the sense under discussion but also in the elementary sense. In this respect my first two illustrations differ from the third, and the contrast helps to bring out the point. For though the number 7 and Hamlet exist in the sense that they are items in their respective sets, we do not in this case assume that they exist in our elementary sense, though philosophers may raise the issue whether they do. Thus someone may without contradiction deny that Hamlet exists on the ground that he is a literary fiction while admitting his existence as a bona fide Shakespearean character.

There are indeed circumstances in which the attribution of existence in two different senses will coincide, and thereby cause confusion. Thus, nominalists, using the word 'exist' in the elementary sense, may assert that only particulars exist and that there are no such things as essences. In doing this, they may appear to be in agreement with those who exclude essences from existence by the verbal conventions they adopt. But we must insist that these are quite different things. While the latter are adopting one usage of 'exist', the nominalists are presupposing another—the elementary one, and on that basis they are presenting us with a metaphysical theory about the kinds of things which exist in their sense. They are like those I have mentioned who, using 'exist' in the elementary sense, assert that to exist is to be perceived or to be spatio-temporal. Their concern is not about the use of the word 'exist' but about the furniture of the world.[3]

It may be objected that I am here drawing an excessively sharp distinction between questions of meaning and questions of fact. There is of course no doubt that these two kinds of questions are intricately related in the construction of theories. But we nevertheless have to recognise the distinction if we are to avoid a very real danger of confusion. Those, for example, who speak of modes of being or ways of being often fail to make it clear whether they are discussing senses of 'being' or kinds of beings. I have been concerned here to do the former, with the purpose of isolating the elementary concept of existence from others for which the same word is used.

THE ELEMENTARY CONCEPT OF EXISTENCE

We need now to say something more about this elementary concept. For, though it is an elementary concept, in the sense that it is not possible to give any reductive account of it, it nevertheless has a content which we need to make explicit. There are, that is to say, certain requirements to which everybody is implicitly committed when using the word 'exist' in its elementary sense. There are some admittedly who apparently do not accept that they are so committed, and why this should be so is something we must consider. It should be said straightaway, however, that in my view they have failed to grasp the concept of existence.

The main requirements which I have in mind are as follows:

(1) ***Either something exists or it does not***. There is, that is to say, no third alternative to existing or not existing. This is no more than to say that in talking of existence in the elementary sense, we are accepting the law of excluded middle. Or, to put it another way, the principle of bivalence holds for all existential statements; they are all either true or false. The traditional reference to excluding the middle suggests that the third alternative would be a sort of compromise, 'hovering somewhere', as Plato expressed it in writing of the sensory world, 'between the realms of non-existence and full existence'.[4] But this is not essential; if something were neither to exist nor fail to exist, nothing need be said about where it hovers. And in fact recent critics of bivalence have envisaged those statements which are neither true nor false as lying in a sphere outside and beyond that in which the dichotomy between truth and falsity applies.[5] What I am here pointing out is that in terms of the elementary concept of existence, there could be no such realm or sphere, wherever one might locate it. For in terms of this concept, it would always be relevant to ask about anything allegedly within that sphere, whether or not it existed, and to expect a yes-or-no answer.

(2) ***There are no degrees of existence***. Nothing, that is to say, can have more or less existence than anything else. The suggestion that it can may be regarded as an alternative to abandoning the law of excluded middle. Even if there is no further possibility beyond existing and not existing, a range of possibilities is opened up within the sphere of existence, some things having a higher grade of being than others. When Plato said that the sensory world lay between non-existence and full existence, the inclusion of 'full' suggests that he was prepared to fall back on the idea that it did after all exist but only on a lower level. Since his time, however, the idea that there are levels or grades of being, has often been asserted in its own right, reaching its most

systematic exposition in the Hegelian doctrine of degrees of truth and of reality. What I am claiming is that existence, in its elementary sense, is not something that there in this way can be more or less of. Some things may be more comprehensive, more stable or more valuable than others, but they cannot be more real.

(3) ***There are no kinds of existence***. This proposition differs from the previous one only in being qualitative rather than quantitative. The basic idea is the same. Instead of saying that something has more of existence than something else, one might say, for example, that it has necessary rather than contingent existence, real rather than actual existence, or existence rather than subsistence. To speak of kinds of existence in this way is, I suggest, to make a conceptual mistake. There are of course different kinds of thing that exist, but that, as I have insisted, is another matter. Descartes envisaged a world in which there were two very different kinds of existent thing—the material and the mental. Those who have attributed to him the view that these had different kinds of existence have attributed to him something which no-one can seriously envisage. This has been notably, though ambiguously, expressed by saying 'there is only one way of being',[6] it being meant by this neither that the word 'being' is univocal nor that everything that is is of the one sort, but rather that being, in the elementary sense, admits of no variety in itself. To exist is to exist without qualification.

(4) ***Existence is not a relative concept***. There is, that is to say, no question of something existing in one context or for one person, while not existing in another context or for another person. What exists is in no way dependent on one's conceptual framework, one's universe of discourse or one's culture. It is not possible, for example, for people with supernatural powers to exist or be a reality for the Azande and not for us.[7] To exist is to exist absolutely.

Another way of putting this is to say there is no possibility of there being different worlds with different kinds of existence. The proposition that existence is not a relative concept can therefore be taken as a special case of the denial that existence can be of different kinds, since existence-for-me can be taken as one kind of existence and existence-for-you as another. When we deny that there could be such different worlds, they must not be envisaged, as it is all too easy to do, as existing alongside one another, as it were, like galaxies in physical space or people with diverse cultures in Central Africa and Europe. For this would be to represent them as being no more than parts of one world, with you and me inhabiting different parts of it. When it is said that one world exists for me and another for you, this is clearly not what is intended. It is a picture which comes to mind solely because what is intended cannot be seriously envisaged.

(5) ***Existence is not a property***. Or perhaps we should express this more cautiously by saying that attributing existence is something completely different from attributing a property. This difference is evidenced by the fact that when we attribute a property to an entity—that is, to an existent thing—we can always at least envisage some other entity which does not have it;[8] whereas we cannot envisage an entity which does not exist. That everything exists is a necessary truth.

This statement should not be confused with the much discussed statement that existence is not a predicate, or with the more refined one that it is only a second level and not a first-level predicate. Whether it is legitimate to predicate existence of anything at the first level is not an issue on which our principle need commit us.[9] All we need do is point out that, even if it is legitimate, predicating existence is a radically different kind of operation from predicating properties. It is uninformative in the sense of adding nothing to the description of the thing; the only information it gives us is that some given description has instances.

(6) ***We cannot classify objects into existent and non-existent.*** We cannot allow that whether an object exists or not is an issue to be resolved when we are describing it. For we could not begin to describe it unless we already treated it as existing. Existence, as I have said, is something that everything has.

It could be argued that this requirement is not independent of the previous one, since it is reasonable to suppose that if existence is taken to be something which belongs to some objects and not others it is being treated as a property. But whether this is so or not it merits separate mention because of the conspicuous way in which it has been challenged by Meinong and by the phenomenologists. To accept it is to claim that an elementary concept of existence rules out both the non-existent objects of Meinong and the intentional objects which Husserl took to be an object of study after excluding their existence.

We should perhaps add to it as a subsidiary requirement that there can be no reference to, or experience of, what does not exist. But we can do this only if reference and experience are treated as relations. If they are relations, and we accept that both the terms of any relation must exist, then reference to or experience of the non-existent is eliminated along with the non-existent. Words like 'reference' and 'experience' are no doubt often used in other ways, but in this case they do not concern us.

THE EXISTENCE PRINCIPLE

These six requirements which I have listed are intended to include at least most of what we have in mind when we speak of existence in the basic sense. The list may well not be exhaustive. No attempt is being made here at architectonic completeness. We might wish, for example, to include the requirement that everything that exists is determinate. This is something we will consider in due course.[10] But we should be careful about where we draw the line. While the law of excluded middle, for example, is a requirement of the concept of existence, the law of non-contradiction is clearly not. If someone says of anything that it both exists and does not most of us would regard this as an absurdity. But anyone who says it has presumably already grasped the concept of existence.[11]

The items which I have listed are, of course, connected in various ways. The first four may be summed up as expressing the idea that all questions about existence, without restriction, require an unqualified yes-or-no answer,[12] there being no room for the existence in question to vary either quantitatively or qualitatively or in relation to context. The last two, though independent of this theme, follow it up with an insistence that saying something exists gives no information about what kind of thing it is, and hence does not differentiate it from anything else. This in its turn is closely connected with saying that the concept of existence is an elementary one, since if existence were to be reduced to something else it could only be to some general property or other.

That there is such an irreducible concept of existence which we constantly employ, and that it has the features which I have set out, is something which I propose to put together under one heading and call 'the existence principle'. It is a principle of a somewhat special kind. It is not an assertion about the world, not even an assertion that there is a world. It is rather an assertion of what is required for there being a world at all. Only when it has been accepted can we go on to ask straightforward metaphysical questions about the content or the character of this world—questions such as 'Do universals exist?', 'Does matter exist?', 'Does God exist?', 'Is everything spatio-temporal?'. When such questions are asked, the principle puts us under an obligation to give an unrestricted, unqualified, yes-or-no answer.[13]

Given its special status, it might be supposed that this is a principle which could not be in dispute, since it is presupposed by all our common thinking about the world. It has however been challenged in various ways and in each of its clauses, sometimes explicitly but more often implicitly, by people who have presented us with theories about the nature of things. In outlining its clauses I have already mentioned some of the forms of this challenge. Something more, however, needs to be said about them if we are to be quite clear why such a basic principle needs defending.

There has, it is true, been little direct challenge to the idea with which I started, that existence is an elementary concept. The trouble in this case is rather that forms of expression are used which leave it unclear whether a reduction is intended or not. To return to my original examples, to say that to exist is to be perceived or to be spatio-temporal is at least to suggest that existence is to be identified with one or another of these general properties of things. In many cases these may be no more than dramatic ways of expressing metaphysical theories about what kinds of thing exist. But even if so, we need to clear up the uncertainty, and insist that they could be nothing more.

There is moreover the case of George Berkeley who appeared to take quite literally his identification of existing with being perceived, at least where 'sensible things' were concerned. We need only consider the meaning of the term 'exist', he said, to see that this was so.[14] And that he was serious about this is borne out by his claiming as a central argument in its favour, that the existence of unperceived things was a manifest contradiction.[15] Here then we have a direct challenge to the elementariness of the concept, and, incidentally, to the view that existence is not a property.

The elementariness of the concept, however, has not been the main point at issue. Even when its elementariness is admitted, there are still those who have denied that the law of excluded middle applies to it, or have dealt in various grades or kinds of being, or have treated existence as a property on other grounds than its reducibility.

Let us take first the requirement that we accept the law of excluded middle. This may well have been already put at risk by Plato when he said that the sensory world hovered between existence and non-existence. For this might be interpreted as an application to existence statements of the general idea that there are borderline cases to which the law of excluded middle does not apply. This is an idea which has had a long history, and which has recently been given standing as valid for an alternative inexact logic of common sense.[16] The challenge to this requirement, however, has become even more conspicuous in the views of mathematical intuitionists, adopted and generalised by Michael Dummett, according to which there are limits outside which the bivalence of truth and falsity does not hold.

The proposition that there are degrees of existence also has a long history, even if only made wholly explicit by Hegel. If Plato granted any kind of reality at all to the world of sense, he certainly thought of it as having less reality than the world of forms. And, after Plato, there was a continuing conception of a hierarchy of being, leading right up to God, the 'ens realissimum', at the top.

From this notion of a gradation of being we can move on very easily to a differentiation of kinds of being. Plato's two worlds were not to be regarded as parts of one world, but rather as worlds distinguished by the kind of existence they had. Meinong's use of the word 'subsistence' for one of these kinds makes no difference to this distinction, though it implies a grading of two kinds which was different from Plato's. The distinction between kinds has also been implied by those who have talked of necessary existence, distinguishing it from contingent existence, the necessity here being held to be 'de re' and having nothing to do with the cogency of the proof of it.

Then there is Kant's distinction between the empirical reality of the world of appearances, and the transcendental reality of things in themselves; F.H. Bradley's distinction between existence, which he assigned to appearances, and reality which he assigned to the Absolute; David Lewis' distinction between the existence of possible worlds and the actuality of one of them. In all this we should not forget Berkeley. In so far as he fell short of identifying the existence of ideas by definition with their being perceived, he seems at least to have been involved with assigning them a special kind of existence, to be distinguished from the existence of spirits.

When we turn to the proposition that existence is relative in character we can start, as everyone knows, with Protagoras who said, according to Plato, that man is the measure of all things, of the things which are, that they are, and of the things which are not, that they are not. Plato's classic criticism of this view has not stopped it from erupting throughout history and from being especially prevalent at the present time. I have taken an illustration of this view from Peter Winch who has claimed according to a general understanding of him, that what was real for the Azande tribe of Central Africa was not real for us who are embedded in a Western scientific culture. But what exists has been said by others to vary not only with one's culture, but with one's scientific paradigm, or even more generally with the categorial scheme which one adopts. This proliferation of relativistic views provides a widespread challenge to the fourth item listed under our existence principle.

A challenge to the fifth item, whereby existence is treated as a property, is basic for the ontological argument for the existence of God and hence is of long standing. It also lurks in the background when people look as if they are identifying existence with some general feature of the world. And finally it is linked with the rejection of the sixth item—a rejection which is exemplified, as I have mentioned, in Meinong's views on non-existent objects, and in the view of the phenomenologists that one can engage in a study of fields of phenomena without admitting their existence.

At this point, mention should be made of Martin Heidegger, who followed in the steps of the phenomenology of Edmund Husserl. Though he said nothing about studying phenomena minus their existence, he envisaged an enquiry into the 'meaning of Being', which put him into clear conflict with other requirements of the existence principle.[17] His concern was not with the meaning of the word 'Being' ('Sein'), nor with an account of what exists (Seiendes). His fundamental ontology therefore seems to involve him in considerations of what I have called a meta-metaphysical kind about the concept of existence. There may be dispute about the significance of what he says on this topic (it is mainly about the meaning of human existence, 'Dasein'), but it is clear that he challenges the elementariness of the concept, and he talks freely of different kinds of Being, such as the authentic and the inauthentic.

Despite these many violations of the existence principle throughout the history of thought, we have had to wait until quite recent times for its formulation and defense. What was needed, it seems, to stimulate an appeal to it was the noting of the surprising and even paradoxical consequence of some of its violations. Thus we may surmise that it was the surprising character of the ontological argument in which the existence of God was proved quite simply from his nature, that led Kant to appeal to an early version of item 5 of the principle for the rejection of the argument. Similarly it was the paradoxical appearance of Meinong's violation of item 6 which led Bertrand Russell to formulate a more sophisticated form of Kant's appeal.[18] And finally it was the oddness of Bradley's assertion that Time existed but was not real which led G.E. Moore to question the whole idea that there could be kinds of existence at all. 'What more', Moore asked wonderingly, 'can a man who asserts that Time is real mean to maintain about it than that it exists?'[19]

Starting from the time when Moore was asking his rhetorical question, there have been two writers, in particular, who have moved towards a general statement of the principle. The first of these is Samuel Alexander who, in the opening pages of his metaphysical treatise 'Space, Time and Deity' affirmed the principle that in respect of being or reality, all existences are on an equal footing.[20] Using an intriguing analogy, he said that as far as existence is concerned, the world is a democracy. He described this as the attitude of realism, but the using of this term may, as we shall see, lead to some confusion.

It was John Anderson, however, author of the papers included as 'Studies in Empirical Philosophy' who has given, as far as I know, the greatest emphasis to the principle, and has been the most systematic in following through its implications.[21] He is the author of the slogan 'There is only one way of being'. This statement may be, as we have seen, ambiguous as it stands, but there is no doubt that he envisaged it as covering not only item 2, but all the clauses I have included in the principle. He could hardly have done this without also accepting that the concept of existence was an elementary one, though unfortunately he was one of those who gave an impression of its reducibility by appearing to identify existing with occurring in space and time.[22]

Anderson referred to this principle as the distinguishing mark of empiricism, and he was prepared to defend his use of this description. We might advisedly dismiss it, however, along with 'realism', as liable to cause confusion. It has subsequently been referred to as 'existence-monism'[23] but this again has its dangers, since monism is normally taken to be the metaphysical theory that existents are all of one kind or form one system, not that existence itself does not admit of kinds, degrees or alternatives. It would be wiser, therefore, to speak, as I have done, of the existence principle and leave realism, empiricism and monism out of it. This will help to remind us that what we are concerned with is not a metaphysical theory about the nature of the world but something which underlies the formulation of any such metaphysical theory.

NOTES

1 Nicholas Rescher, 'The Equivocality of Existence', in *Studies in Ontology* (American Philosophical Quarterly Monograph Series, Vol. 12, Oxford, 1978) Sect. V, pp. 62-64, refers to existence in this sense as 'genus-relative existence'.

2 See John Austin, *Sense and Sensibilia* (Oxford, 1962) Ch. 7, where much is made of such diverse uses of 'real'.

3 Nicholas Rescher, *op. cit.*, unfortunately does not make this distinction when commenting on nominalism. See p. 66.

4 Plato, *Republic* M.P. 479. The translation is that of Lee in the Penguin edition.

5 See e.g. Michael Dummett, *Truth and Other Enigmas*, pp. XXX and elsewhere.

6 See John Anderson, *Studies in Empirical Philosophy* (Angus & Robertson, 1962), p. 3.

7 See Peter Winch, 'Understanding a Primitive Society', in *American Philosophical Quarterly*, Vol. I (1964) for what is real for the Azande.

8 I exclude mention here of those formal properties which necessarily belong to everything that exists, such as that of being identical with itself. These, of course, are entailed by the fact of their existence.

9 For some recent differing views in this question see J.L. Mackie, 'The Riddle of Existence', in *Aristotelian Society Supplementary*, Vol. L (1976) pp. 247-265; C.J.F. Williams, *What is Existence?* (Oxford, 1981); Barry Miller, '"Exists" and Existence', *Review of Metaphysics*, Vol. XL, No. 2 (1986) pp. 237-270.

10 See Ch. 6, p. 101 *et seq.*

11 For more on contradiction, see Ch. 5, p. 82.

12 Unless of course we include the possibility of contradictions in reality and add as an alternative the answer—yes *and* no.

13 The term 'metaphysics' can of course be used to cover more than such straightforward metaphysical questions. In particular it can be used to cover the questions about the concept of existence with which we are here concerned. In what follows, however, I use it in the restricted sense in which its concern is simply with what there is.

14 George Berkeley, *The Principles of Human Knowledge*, Sect. 3.

15 *Op. cit.*, Sect. 4.

16 See, e.g. Stephan Körner, *Categorial Frameworks* (Oxford, 1970).

17 See M. Heidegger, *Sein und Zeit* (7th ed., Tübingen, 1953) translated as *Being and Time* by John Macquarrie and Edward Robinson (London S.C.M. Press, 1962).

18 See Bertrand Russell, 'Meinong's Theory of Complexes and Assumptions', *Mind* (1904) and 'On Denoting', *Mind* (1905).

19 See the paper 'The Conception of Reality' in G.E. Moore, *Philosophical Studies* (1922) Essay VI, p. 199.

20 S. Alexander, *Space, Time and Deity* (1920) Vol. I, p. 6.

21 John Anderson, *Studies in Empirical Philosophy*. See esp. the first paper, entitled 'Empiricism' (1927), pp. 3-6.

22 See Anderson, *op. cit.*, p. 14.

23 See John Passmore, *Philosophical Reasoning* (1961) pp. 39 *et seq.*

CHAPTER TWO

THE DEFENSE OF THE EXISTENCE PRINCIPLE

What then are we to say to all those who have violated the existence principle in such diverse ways?

THE APPEAL TO MISUNDERSTANDING

The first move that can be made in its defence is to suggest that the violations are only apparent, and that the dispute can be resolved by a more careful consideration of the uses of such words as 'exist' and 'real'. Those who reject the principle, we may say, are still being influenced by other uses of these words in a way of which they are unaware. Misunderstandings so created, we could point out, are all the easier where problems of translation are involved—if, for example, you are an English-speaker and they write in ancient Greek, using words that in general correspond to 'exist' and 'real'.

To argue in this way is of course to appeal to consensus, and it does not follow from an undisputed acceptance of the principle that it is correct. In these circumstances, however, it is a considerable point in its favour if no-one doubts it. And there are at least some cases where the argument that the apparent dissidents do not really doubt it can be quite effective.

Consider, for example, some of the contexts in which people have talked of degrees of reality. It may be suggested that when Plato distinguished (in Greek) between the full reality of the forms and the shadow reality of the world of sense, what he really had in mind was that the latter was dependent on the former. Grades of dependence can be easily envisaged. When God was said to be the *ens realissimum* having the maximum degree of existence, one explanation of this could be that He was conceived of as the most independent of all entities. In the same way, material things may be thought of as more real than castles in the air because they are not dependent on us. Realities are in general thought of as what we cannot do anything about; they are harsh and have to be faced. When Freud distinguished between the reality principle and the pleasure principle, it was presumably such reality as independence that he had in mind. This clearly has nothing to do with existence in its elementary sense.

For a second suggestion, we may go back again to Plato. Something that he probably had even more in mind than independence was that the forms were *unchanging*. To be transient and fleeting, it may be thought, is to lower one's quota of existence. There is indeed some point in this. The longer anything endures, the more time it has to exist in, and therefore the more existence it may be supposed to have. But this of course is to confuse the amount of what exists with the amount of its existence. We might as well say, on these lines, that there is more existence in an elephant than in a mouse. Mere extent, whether in space or time, cannot confer a higher grade of existence on anything.

Plato of course was one of those who regarded any change as a bad thing. This suggests another element involved in conferring on his forms a higher grade of existence. The forms were eternal, existing beyond the flux of time. They were therefore simply of *greater value*, and in that sense a superior form of existence. There is no doubt that the worth-whileness of things is a matter of degree. So the assimilation of reality to value is sufficient to account for the hierarchy of being. Existence as independence may on occasions be unpleasant and harsh, but this is more than compensated for by existence as goodness. God as the *ens realissimum* is not only the most independent of beings but the one which is supremely good.

To speak of something as the most real of entities has also other implications. For the Leibnizians, and for Kant who followed them, it was to speak of that which possessed all positive attributes in a maximum degree. Once again, it is the possession of *more* of something which is supposed to boost the degree of reality. This idea is manifested in its most straightforward form in the Hegelian notion that the whole is more real than its parts. By definition, of course, the whole is greater than its parts. But the transition from 'is greater than' to 'has greater reality than' remains a plain non-sequitur. And an exactly similar non-sequitur is manifested in the Leibnizian transition from 'having more attributes than' to 'having more reality than'.

When we speak of degrees of reality, then, we may well have in mind various ways of grading things in the world into different sorts of hierarchy, in terms of independence and power, in terms of value, in terms of extent and of various forms of comprehensiveness. But it does not follow from this that there is a hierarchy of existence in its elementary sense. It can still be the case that from the point of view of such existence, the universe, as Samuel Alexander said, is a democracy, all existents being on an equal footing.

The case of degrees of reality is not the only one in which departures from the existence principle may be more apparent than real. I have already mentioned the inclination of some to speak as if existing could be reduced to some general feature of the world (such as being spatio-temporal) when all they in fact have in mind is a metaphysical theory to the effect that everything which exists has this feature. Then there are those apparent believers in relative existence, who in fact have in mind something much more innocuous. When Thomas Kuhn said that after a scientific revolution scientists are in a different world he may have meant no more than what he had stated a few lines previously and embodied in his chapter heading,[1] namely, that after a revolution they see the world differently. Again, when G.E.M. Anscombe says that there are things which do not exist and yet have properties, mentioning her great-grandfather, who is a grandfather, and Protagoras' Treatise on Truth, which is famous, we are to assume that she is not challenging the sixth item of our principle, but rather focussing attention on a use of 'exists' which excludes things past which have 'ceased to exist'.[2]

It would be nice if all cases of departure from the existence principle could be handled in this way. We could then simply rely on a consensus and forget about the issue. Unfortunately this would be to ignore a certain motivation for departing from the principle which goes well beyond any verbal misunderstanding or any delight in using honorific words like ‘real’. When the question is raised whether things of certain kinds exist, there may be a quite genuine and serious problem about how we should answer. When the distinction is made between appearances and the realities of which they are the appearances and we are asked whether the appearances exist, we are inclined to say on the one hand that of course there are such things as appearances, and on the other hand that this cannot be so because we have just distinguished them from the realities. Again, when we find that among the Azande there is a belief in the existence of witches, and they operate within a framework of concepts which is incommensurable with our own, have we the right to deny this, thereby imposing our ‘Western scientific’ ways of thinking on them? Then there are such questions as whether the future exists as well as the present or the past, whether properties exist as well as the things which have them, whether what is possible exists as well as what is actual. In all such cases, we may find it difficult to say yes or no, and this will lead us to the thought that we can solve the problem by making a distinction, and saying that there are perhaps different kinds of existence, or that existence is relative to a framework, or that there are some cases where existence claims are irrelevant, or that there are some things which have a status as objects, but which do not exist. Hence we envisage distinctions between real existents and apparent existents, between real objects and intentional objects, between things which exist for some and things which exist for others, between substances which exist and properties which subsist, between present events which exist and future ones which neither exist nor fail to exist, between the actual world and possible worlds.

It should not be denied in such cases that there is a genuine problem about what we should say. But given an acceptance of the existence principle, this way of resolving the problem can only be described as hedging. The plain fact, we might say, is that a great many philosophers throughout the ages, when faced with straight metaphysical issues, have preferred to evade them, rather than attempt to settle them, and the proliferation of departures from the existence principle has been a primary device for evasion. What we should do is rule such evasions out from the outset and demand in each case a yes-or-no answer.

The making of such charges does of course presuppose acceptance of the principle. So we are still left with the need to justify its acceptance, and, we must now add, to provide a justification for it which is sufficiently strong to offset what would otherwise be an easy way of resolving some difficult problems.

THE APPEAL TO LOGIC

It must be admitted straightaway that it is not possible to justify the principle empirically. We have various ways of establishing that things exist, by observation, for example, as with the continent of America, or by inference, as with the planet Neptune. But if claims are made that some of these things have more existence than others, or existence of a different kind, there is nothing further we can point to to prove them wrong. There is also of course nothing further that the makers of such claims can point to to prove them right. Existence claims may well be decided empirically, but the nature of an existence claim cannot.

In these circumstances, the most obvious thing to do is to say that the issue, since it cannot be decided empirically, is a matter of logic. Departure from the principle, it may be said, has to be rejected because it is in breach of some logical law. This would be particularly persuasive if the law in question were the principle of non-contradiction since of all logical laws this is the most difficult one on which to contemplate throwing doubt.

It was in fact an appeal to this principle which G.E. Moore used against Bradley in his paper on 'The Conception of Reality'. His main argument was that Bradley was contradicting himself when on the one hand he said that Time was not real but only existed, and, on the other hand he admitted that what existed belonged to reality. How, said Moore, could anything belong to reality without being real?

The trouble with this argument is that Bradley did believe that something could belong to reality without being real. He did this because he was working with the Hegelian idea that the whole is more real than its parts. If you accept this idea, parts, such as he considered Time to be, could belong to the whole and still only exist, even though only the whole is real, in the sense of completely real. Moore, of course, did not accept the notion that there could be such degrees of reality, and hence failed to appreciate that anyone who did accept it could at least avoid the contradiction. I say 'at least' because this would not free Bradley from the suggestion I have already mentioned that when he spoke of degrees of reality what he really had in mind was degrees of comprehensiveness. The point is rather that if he refused to accept this suggestion, and stuck to the idea that there were degrees of reality or of existence in its elementary sense, determined by degrees of comprehensiveness but not identical with them, there is nothing further, on this line of argument, which could be done about it.

In the same way it is hard to see how any other departure from the existence principle could be shown to be self-contradictory unless one already accepted the principle. There is another obvious illustration of this when it is said that the belief in relative existence involves its holders in contradiction, since they are prepared to allow two incompatible statements about what exists both to be true. The answer to this, of course, is that so long as you stick to the relativity of existence statements, the two statements are not incompatible, being about different worlds.

Following G.E. Moore, John Anderson was another who defended his principle that there is only one way of being by an appeal to logic.

In his case, however, the appeal was not to the law of non-contradiction but to the law of excluded middle. It was a matter of logic, he said, that there could be no alternative to being or not being.[3] However we can see at once that there is a special trouble with the appeal to logic in this case. The law of excluded middle is part of the existence principle itself, and anyone who rejects it as part of that principle will of course also reject it as a law of logic. Furthermore its status as a logical law has itself had doubt thrown on it, by mathematical intuitionists and others. Hence any one who defends it as part of the existence principle would be advised to do it directly, and not by appeal to it as a logical law.

We will need to say more about the law of excluded middle in due course.[4] But the fact that we allow it a dual role as part of the existence principle and as a principle of logic (though one under challenge) suggests a need to say something in general about the relation of logic to the concept of existence. We should begin by distinguishing between two different views about the nature of logic. There is one view of logic which makes the choice between alternative systems a matter of convenience. There is more to be said about this,[5] but if it is taken at its face value, support for any logical system will be independent of any considerations about what exists in the world, and a fortiori about the concept of existence. On such a view therefore we can keep the law of excluded middle as part of our principle, whatever system of logic we find it convenient to use.

There is however another view of logic according to which its principles are very general statements about the structure of the world. On this view, a 'realist' view we might call it, its relation to the existence principle is more complex. Its principles are in a very broad sense metaphysical principles; presupposing that there is an existent world they say something in very general terms about what it is like. This means that they say nothing about the concept of existence itself. They are concerned with the nature of the world, not with what is required for there to be a world. This is why we can speak of the law of excluded middle, if we accept it as a law of logic, as having a dual role, one which is not possessed by other laws of logic, such as the law of non-contradiction, which are not requirements of the concept of existence.

It will be clear from this why the appeal to logic will be ineffective in this case. Our concern is not merely that the law be accepted, it is also that it fulfils its role as part of the existence principle. To establish it as a logical law would meet the first concern, but not the second. It would say something about the limits of possible existence, but nothing about whether it is one of the requirements for anything to exist at all. So appealing to its status as a law of logic would leave us where we were.

At this point, we may well feel inclined to stop beating about the bush and to reiterate that those who depart from the principle have simply failed to grasp the concept of existence. All we can do is try to get them to grasp it, perhaps by asking them to tell us more precisely what they have in mind. Take, for example, Kant's distinction between kinds of existence, one possessed by the empirical world, and one by the transcendental world. We have various ways of distinguishing between these worlds of his; one is known, the other not, one is subject to certain categorial laws, the other perhaps not. What more do we add to such distinctions, we may ask, by saying that one has empirical existence the other transcendental existence? Is it not clear that all Kant was doing was to posit that both simply exist? But if they both simply exist, they can no longer be said to be different worlds, but only different parts of one world. Anything more cannot be seriously envisaged. We must seek to exhibit the fact that it is unintelligible, or, if you like, a conceptual mistake. When those concerned have seen that it is so, the argument is finished.

In the end, it seems to me, those who defend the existence principle may have to fall back on to this kind of direct appeal. But everybody knows what the trouble with it is. If you appeal to others to see that what you say is right, they may simply reverse the appeal. If you say that what they say is unintelligible, they may say that this merely shows up your psychological limitations. You are restricting yourself to a too narrow system of categories, and what you need to do is to overcome this, and then you will realise what they are talking about. If you reply that your system is implicitly accepted in the ordinary thinking of most people, they may say that such matters are not to be decided by appeal to the unreflective multitude.

THE APPEAL TO UNACCEPTABLE CONSEQUENCES

Fortunately the matter does not quite end there, with your intuitions being met by counter-intuitions. What we are concerned with after all is a very basic principle, the acceptance or rejection of which can have very widespread consequences. If it can be shown that the rejection of the principle has consequences which those who reject it find as unacceptable as you do, this is clearly an argument in its favour. And the argument is the stronger the more grossly unacceptable the consequences. The argument is of course based on a demand for consistency, and if your opponent is prepared to retain consistency and accept the consequences, it fails. It is a kind of a threat, and one hopes that those to whom it is issued will take heed.

Arguments of this kind have been called transcendental arguments, following Kant who used one to defend certain principles as true of his empirical world. If you didn't accept these principles, he said, you would have to admit that you could have no experience of this world at all, and he assumed that you would not accept this consequence. What I am now suggesting is that an argument of this kind can be used to defend the existence principle. This is something which Kant himself certainly did not envisage involving as it does a rejection of his central distinction between an empirical world and a transcendent world of things-in-themselves.

John Anderson in fact produced a version of such an argument which he took to be closely connected with his appeal to logic. The postulation of different orders of being, he said, was contrary to the very nature and possibility of discourse.[6] What he had in mind would, I think, have been made clearer if instead of discourse he had spoken of the process of enquiry. In any enquiry whatsoever, according to this argument, what we are doing is asking the question 'Is it or is it not so?'. When an issue is raised about what exists in the world, the only way to settle it is to say that it does or that it does not exist. No room is left here for the possibility of its neither existing nor not existing or of its having one kind or grade of existing rather than another, or for its existing for someone and not someone else, or for its being an object of the non-existent kind. To admit such possibilities is therefore to close down on the enquiry.

This consequence could, I think, be fairly considered unacceptable. It is not admittedly as radical as Anderson made it appear when he spoke of the impossibility of discourse.[7] It falls short of total scepticism. What is insisted upon rather is that once you depart from the existence principle, the scope of any enquiry becomes restricted. You can find out, for example, about what happens in Kant's empirical world, and the increase of one's knowledge of it can continue indefinitely. But you could do so only on the assumption that it is the only world, and you could not envisage there being anything apart from it. If you were to envisage two kinds of existence, or even two grades of existence, you could not possibly pursue your enquiries about both, as it were, in tandem, because there would be no way of distinguishing between the two kinds in the course of your dual enquiries. Similarly you could never recognise a phenomenological enquiry about non-existent objects for what it was and would be unaware of any sphere in which the law of excluded middle did not hold.

In reply to this it might of course be said that this consequence does not follow, since in gaining knowledge we are able to recognise the different kinds of existence belonging to what we are gaining knowledge of, and the limitations to which such existence is subject. The trouble, it may once again be said, lies in the psychological limitations of the critic. So here we are back again to the level of assertion and counter-assertion. However, at the very least, the issue has been broadened. It is now seen to concern not only the concept of existence but also the concept of knowledge. And with every extension of their commitments the opponents of the existence principle will find their position harder to defend.

These considerations apply also to another unacceptable consequence which may be claimed to follow, in particular, from accepting different kinds of existence and hence the separation of entities into different worlds. This is the consequence that there could be no relation between the worlds or what they contain. For a state of affairs in which something with one kind of existence was related in any way to something with another would be in a kind of no-man's land between them, and could not be incorporated into either world. What kind of ontological status, then, could it possibly have?[8]

This is a consequence which would be definitely unacceptable to most of those who have envisaged what we might call an existence dualism. Plato's theory of forms allowed for participation of the forms of things in the world of sense. Though God, in the hierarchy of being, was more real than his creatures, they were the result of His creation. Hegel's Absolute was related to things of a lower order as a whole to its parts. And throughout there has been an assumption of some relation of dependence which as we shall see, has even been made the identifying characteristic of a lower order of existence. If all this was abandoned and an entire separation was insisted upon, such dualisms would have little left to recommend them.

Such existence dualism must of course be distinguished from ordinary metaphysical dualism which envisages only two very different kinds of thing in the one world. A metaphysical dualism would in no way be affected by the present argument. It may admittedly have problems about the relation between two very different kinds of thing. If one is eternal, like God, and the other temporal, or if the one is non-spatial, like a Cartesian mind and the other spatial, one may be puzzled about what kind of relation, or causal relation at least, there could be between the two. But these would be internal metaphysical problems, which would arise only after a single ontological status for both elements in the dualism was accepted.

Descartes' dualism of the material and the mental presents us, in this regard, with an interesting case. There is good reason to believe that Descartes was basically a straightforward metaphysical dualist who accepted the existence principle. It must be admitted that in his proofs of the existence of God he made great play with there being degrees of reality and with a distinction between objective (representative) and formal (inherent) reality. But in his general metaphysical speculations he appeared to take it as his job simply to find out what the world was like. It could be that it was for this reason that he saw no difficulty in maintaining that there was interaction between his two substances. His successors, however, did see such a difficulty, and this appeared to have a deeper source than the problem of the non-spatial interacting with the spatial. This may lead us to suspect that they were interpreting his theory as an existence dualism, and appreciating, on the lines of our present argument, that on this interpretation, interaction would be impossible. Given existence dualism, in fact, parallelism would have a lot to be said for it. If this was their reason for maintaining it, however, it must be admitted that it would make them appear much more sensitive to the force of the argument than those many existence dualists who have allowed all sorts of relations between different worlds.

At this point, such existence dualists may once again reply that they see no difficulty in there being relations between things which have different kinds of existence, and that failure to appreciate this special kind of relation is due merely to the limitations of the critic. And once again, all that can be said to this is that if those who depart from the existence principle continue to extend their commitments in this way, their position becomes progressively harder to maintain. Arguments to unacceptable consequences therefore have a place in bolstering up the basic assertion that such departures involve a conceptual mistake.

Having said this much in general terms about the existence principle and its defense, what we need to do now is consider in more detail various ways in which it has been departed from and various consequences of retaining it. This in its turn will also help us to expand on its defense.

NOTES

1 Thomas Kuhn, *The Structure of Scientific Revolutions*, Ch. X, p. 111, 2nd ed. (1970).

2 G.E.M. Anscombe, 'Existence and Truth', in *Aristotelian Society Proceedings* (1987-8) pp. 8-10. I neglect the extension of her theme on pp. 11-12 which raises other issues.

3 John Anderson, *Studies in Empirical Philosophy*, Ch. 1, pp. 4-5.

4 See Ch. 3, p. 37 *et seq.*, Ch. 6.

5 See Ch. 6, pp. 93-94.

6 John Anderson, *op. cit.*, p. 4.

7 Anderson also had the idea, which we are excluding, that the use of 'is' as copula in any proposition presupposed that there was only one way of being and that all discourse involved propositions.

8 See John Passmore, *Philosophical Reasoning*, Ch. 3, entitled 'The Two-Worlds Argument' for the presentation of considerations of this kind.

CHAPTER THREE

REALISM AND THE EXISTENCE PRINCIPLE

Since the existence principle is not part of any metaphysical theory about the world it should be impartial when it comes to deciding which of two opposed metaphysical theories to accept as an account of what the world is like. If someone says the world consists of minds or of experience, and someone else denies it, the dispute is about what the world is like, and on this the principle as I have outlined it has nothing to say. In so far then as the traditional dispute between idealists and realists can be characterised in this way, there is no inconsistency between the existence principle and an idealist metaphysic.

Why then, we may ask, did Samuel Alexander describe the principle as he presented it, as the principle of realism? And why has the recent rejection of the law of excluded middle, and therefore of the principle, been described by those who have made it, as anti-realist? The answer obviously must be that the traditional dispute has elements within it which make it something more than the straightforward metaphysical dispute which I have described.

In the case of Alexander, it might be thought enough to point out that the idealism which he opposed was the absolute idealism of Bradley and of Bosanquet, and that this included within it not only the idea that reality consisted of mind or of experience, but also the idea that the whole was more real than any of its parts. It was in this second respect, it might be said, that it constituted a direct challenge to the principle, but since this was so much bound up with the idealism of the day, it is not surprising that those who rejected this grading of reality would regard their rejection as a key element in their realist philosophy.

KNOWLEDGE-DEPENDENT EXISTENCE

This, however, it not sufficient as a general account of the matter. There is something more deep-rooted in any idealist account of things which has put it at loggerheads with the existence principle. It has to do with the status of the objects of knowledge.

There are in fact two distinct theses which commonly go under the heading of idealism. One is that the objects of knowledge are mental in character. When Berkeley said that they were 'ideas in the mind' this is what he appeared to be saying. If he had just left it at that, we might say that he was accepting the existence principle, and presenting us with a straightforward theory about the character of what exists.

It is extremely doubtful, however, whether this is what Berkeley was primarily concerned with. For him, it might be said, it did not really matter what sorts of things ideas were. The spatial metaphor, 'in the mind' might suggest that they were parts of minds, and shared their character. But he certainly came to interpret the metaphor rather loosely, and to envisage ideas and spirits as distinct entities of which one had different kinds of knowledge. What he continued to insist on was only that ideas would not exist if they were not known. This is where we come to the second thesis of idealism. There is a move here from the notion of a mental world to that of a mind-dependent world.[1] And it is when this move is made that conflict arises with the existence principle. To see this we need to be clear about what is involved in the idea of something being dependent on something else.

It may be said that this is obvious enough. To say that something is dependent on something else is to say that that something else is a necessary condition of it, to simply assert, in other words, the counterfactual that if it had not existed, the thing in question would not have existed. And dependence so characterised, it may be pointed out, is in no way in conflict with the existence principle. Both items in the dependence relation may exist in the plain straightforward way specified in the principle.

When we talk of dependence, however, we are inclined to mean something more than this. We want to know what justifies us in asserting such a counterfactual. We want to know what the state of affairs is, such that the counterfactual comes out true. And the most obvious candidate for this state of affairs is there being some *causal* relation between the entities concerned. When we think of dependence, in fact, we are likely to think straightaway in terms of causal dependence. And this would have the advantage that there would still be no conflict with the existence principle. When one existent thing or state of affairs is the cause of another, both cause and effect can exist in the simple elementary sense.

Yet it is surely the case that those who speak of the objects of knowledge as mind-dependent cannot be thinking of a causal dependence of any kind. There are, of course, states of mind which cause things to exist, but perceptions and beliefs, whether or not they have the status of knowledge, are not among them. We know very well that what produces effects on objects are desires, motives, intentions and such like. It is these which are followed by actions, and it is through action that things and states of affairs are brought into existence. This, which might be called creation, is not the work of knowledge. Thinking about something, unfortunately, does not make it so.

This apparently obvious point is, I think, unaffected by the sort of view we take of causation—whether, for example, we take a sparse Humean view of it as regularity of sequence, or think of the causal relation as involving something more. If we do think of it as something more, saying, for example, that one thing acting on another, or bringing something about, cannot be reduced to Humean terms, the point is made to look especially obvious. This, of course, is what Berkeley did, and it certainly never occurred to him that an idea being in my mind meant that I was the cause of it. Admittedly I was the cause of ideas of imagination, but that was because they were under my control, not because I had knowledge of them. All the rest had to be caused by something else, which, for him, had to be God. For the representationalists he opposed, it was of course independently existing things which did the causing. These, as Locke put it, had the power to produce ideas in us. For neither party could their being objects of knowledge have had anything to do with it.

If we take a minimal Hume-type view of causation, this lack of causal connection does not admittedly look so obvious. This is particularly so when we remember that dependence requires only necessary conditions, not sufficient ones. On this view, all that is required is that there be a generalisation connecting state of mind with object. And the generalisation need not be to the effect that whenever there is such and such a state of mind there is an object, but merely that whenever there is an object, there is a state of mind.

There will be some, no doubt, who will say that this shows how wrong the Hume-type view of causation is. However, it should be remembered that no sophisticated modern Humean accepts that every generalisation supports counterfactuals, or implies a causal connection. What is needed for this is that it be part of a general theory, and in this case there is no sign of any theoretical backing for the proposition that whenever anything exists, some state of knowledge of it also exists. On the contrary, we have a well-established general theory of a folk-psychological and physiological sort about how we interact with the world, and this theory leaves no room whatever for the generalisation in question. It does not allow, in short, for knowing to be a kind of creating.

There is admittedly no direct inductive evidence for the existence of unknown objects, and hence the hypothesis of the causal dependence of objects on the knowledge of them cannot be challenged by appeal to negative instances. But this simple and obvious fact has itself a simple and obvious explanation; what has been called the 'ego-centric predicament' of the knower makes such direct evidence impossible. Hence appeal to this fact can hardly be pitted against a very well-established theory about how we interact with the world.

How then are we to understand the idea of the existence of anything being dependent on its being known? As an alternative to causal dependence we might think of logical dependence, and this, no doubt, is what Berkeley had in mind when he made a move in the direction of *defining* the existence of objects in terms of their being known. If this were to be taken seriously, it would indeed be in breach of the basic idea of the existence principle that existence was an elementary concept, since it would be treating existence, in the case of objects, as a relational property of being known. But it is doubtful whether it could be taken seriously, or even whether Berkeley so took it, for all his talk of the self-contradictoriness of unknown objects. For one thing, it could be applied only to the existence of ideas and not the existence of spirits. This would imply that the word 'exist' was to be treated as ambiguous, having an elementary meaning when applied to spirits, and a reducible one when applied to ideas. Since there is no reason whatsoever to suppose such an ambiguity, this suggestion that the dependence on knowledge is a logical one would be very hard to maintain with any consistency.

If we reject both causal and logical dependence on knowledge, we need to ask what other kinds of dependence there could be of the known on the knower. Might there not be other kinds and might they not fit this case?

It has been said, for example, that there are certain dependency relations between events which are certainly not causal. Xanthippe's becoming a widow is dependent on the death of Socrates, and the whole of a liquid's turning yellow is dependent on a part of it doing so.[2] That there is no causal dependence in such cases may well be agreed. But it could well be argued that such dependence is broadly logical, and therefore not a genuine alternative at all.

Another example which might be claimed is that of the dependency relation involved in what has been called 'supervenience', as in the supervenience of the mental on the physical. May not this be taken as a *sui generis* type of dependence? There is of course no doubt that it fulfils the requirement that the one item is a necessary condition of the other. But those who have looked for a ground for this relation have once again fallen back either on saying it is causal or that it is logical.

We will be saying more about other possible cases later.[3] But for the present it looks as if any attempt to broaden the possibilities of dependence between ordinarily existing items is going to draw a blank. And if this is true in general, it applies all the more forcefully to the dependence of objects on knowledge, which fits none even of the suggested kinds. What then could Berkeley have had in mind?

We have been speaking here throughout of a relation of dependence between items which exist, exist, that is to say, in our elementary sense. If we stick to this condition, there seems to be no alternative left. This brings us, however, to a further possibility. This is that we could still admit dependency of a quite different kind if we abandon the condition. This is a final alternative to which we may be driven. It is one suggested by the use of the phrase 'exists for' when it is said that something exists for something else. Objects which exist for a mind will be dependent on the mind in that they would not exist without it. But the dependence is, as it were, being written in to the existence itself. It is simply descriptive of the kind of existence which the objects have. It is a special kind of existence which we may call 'dependent existence', and if we want to describe the dependence, we may call it neither causal nor logical but 'existential' dependence.

Whether they realise it or not, idealists in general have, I think, been forced into the adoption of this kind of account of the notion of mind-dependence. It is only this which could be behind the special tie they envisage between existing and being known. And it is in so far as they adopt this account of mind-dependence that they come into conflict with the existence principle. They do so moreover while ignoring the suggestion that distinguishing kinds of existence precludes any relation between them. They are involved in fact in a view of mind-dependence which makes the relation of something which has one kind of existence to that which has another kind into an identifying characteristic of the kind.

It will now be clear why realism, as it is commonly understood, has been more than a straightforward metaphysical theory. Its main point has always been to challenge the idealist theory of the mind-dependence of objects. And inasmuch as that theory involves a rejection of the existence principle, its main point has been to defend the application of that principle in this particular case.

We may call this, if we like, a minimal realism. Its point is simply to reject any tie of this special kind between existence and knowledge. In doing this it says nothing about what kinds of thing exist. It was the sort of realism argued for by G.E. Moore, who admitted it left open the possibility that reality was spiritual.[4] It is possible of course that one may wish to add to one's account of realism, by including the idea, say, that most objects do not have mental characteristics.[5] But this is to introduce a metaphysical component into a theory which has as its basis no more than an insistence that existing is one thing and being known quite another.

The tie between existing and being known has been here envisaged as a tight one. It follows the line of Berkeley's views about perception, according to which the existence of objects was represented as dependent on their being actually perceived. Even Berkeley however did not stick to this. After saying that the existence of his table depended on his seeing and feeling it, he added that if he were out of his study he would say it existed, meaning thereby that if he were in his study he would perceive it.[6] He admitted the motion of the Earth, even though it was not actually perceived, because, he said, if we had been far enough out in space we would have perceived it.[7] Clearly this invites a loosening-up move from dependence on actual knowledge to dependence on possible knowledge.

This was a move which was made by logical positivists when they announced their verification principle. According to this, existence depended on the possibility of its assertion being verified. There must, that is to say, be some way of getting to know of it even if you don't actually know of it. If that way were followed, it might of course be found that the alleged object did not exist, but the issue as to whether it existed or not would depend on its verifiability. The unverifiable could therefore be said neither to exist nor not exist.

Though there is much loosening-up here of the idea of knowledge-dependent existence, it must be insisted that that idea is still there. The whole point, after all, of the verification principle is to tie considerations about what exists to considerations about what is known. According to this principle, we are entitled to talk of the existence of flowers on an unclimbed mountain, only because we can argue for or against it from analogy, or devise means for climbing the mountain, or inspect them by helicopter, or think up some other possible way of finding out about them.[8] We cannot therefore answer the straightforward question whether they exist or not; we can only answer the question on condition that a situation obtains which makes reference to our knowledge. So a relation to knowledge remains as a necessary condition of existence, even though it is also a necessary condition of non-existence. To that extent, those who have insisted that Mach and those who came after him were really idealists in disguise, have had an important point to make.[9]

There is of course a difference between logical positivism and traditional idealism. For people like Berkeley, claims for the existence of unknowable things were simply false, whereas for the logical positivists they were meaningless. But not too much should be made of this difference. It derives from the positivists having a new theory about meaning. For the traditional idealist, the meaning of statements about unknowable things is determined by their truth conditions, that is by whether they are either true or false. For the positivist, the meaning of such statements is determined by the conditions for their justification, that is by whether or not there is some way of getting to know their truth or falsity. So all that happens when you turn to a verificationist theory of meaning is that you make the meaning as well as the truth of existential statements dependent on the possibility of their being known. It follows that, in respect of this difference, positivists are even more deeply involved than traditional idealists in breaching the existence principle.

KNOWLEDGE-DEPENDENCE AND THE LAW OF EXCLUDED MIDDLE

So far, the breach of the principle we have been concerned with is a breach in respect of the third clause in our original list. It consists in the allowing of a special kind of existence, which belongs to certain things and not others. But when we come to the logical positivists it looks as if there is also a breach of the principle in respect of the first clause. For what they allege to be dependent on the possibility of knowledge is not simply the existence of objects but also the whole issue as to whether they exist or not. They allow, that is to say, that the law of excluded middle only applies within certain limits—the limits of possible knowledge. Outside those limits nothing can be said either to exist or not to exist.

To this, logical positivists would of course reply that they have no intention of breaching the law of excluded middle. According to them anything whatsoever that can be intelligibly talked about can be said either to exist or not. Every empirically verifiable statement is for them either true or false, and anything else is not a statement at all, but merely words in grammatical sequence. There are therefore no limits outside which anything can be said neither to exist or not exist.

Given their theory of meaning, this reply is undoubtedly in order, and they can be exonerated from breaching the existence principle in this respect. What would be needed for the bringing of rejection of excluded middle into the picture would be the combining of some form of the verification principle with a theory of meaning which allowed meaning to statements about the unknowable. We could then have genuine statements about what is outside the limits. They would therefore neither be true nor false, and concern themselves with what neither existed nor failed to exist.

The path I outline here appears to be that taken by those 'anti-realists' who follow in the steps of Michael Dummett and the mathematical intuitionists. Dummett claims to adopt a Wittgensteinian theory of meaning as use. How this allows for meaningful statements about the unknowable is not our concern here.[10] But in view of the importance he assigns to rejection of bivalence, he clearly has to allow for it. Propositions are only true or false, on his view, if they are effectively decidable; outside those limits the third alternative holds.[11] The substitution of 'effectively decidable' for 'verifiable' suggests much loosening up of the theory of meaning, allowing it to be applied to mathematical statements and allowing for more flexibility about what is effective. But it retains as a basic idea the knowledge-dependence of truth and hence, it would appear, the knowledge-dependence of existence.[12]

In view of this, it may seem surprising that Dummett leaves us at least with the impression that acceptance of bivalence is in itself the distinguishing mark of realism and that anti-realism consists in its denial.[13] This is clearly not the case. A traditional anti-realist like Berkeley never thought of denying that any statement about what the world contained must be either true or false, and even the positivists, suitably interpreted, did not deny it. On the other hand, there have been various reasons offered for abandoning this view which have nothing to do with the belief in the mind-dependence of all objects of experience. There are some, for example, who would abandon the law of excluded middle on the ground of some kind of alleged indeterminacy in the world itself, as found in the vagueness of its properties, or its future, or its quantum structure. The two distinct breaches of our existence principle, therefore, should not be run together under a single heading such as that of 'anti-realism'. The connection comes about because the idea of knowledge-dependent existence leaves a vacuum where before there was a sphere of the existent but unknowable. If you do not want to deny the existence of such a sphere, as Berkeley did, or say with the positivists that claims to the filling of it are meaningless, the idea of a third alternative to existing or not-existing may well be of some help in filling the gap.

What must now be questioned, however, is whether the rejection of bivalence, as understood by anti-realists, helps in any way in the filling of this gap. To see this we need to look more closely at the relation between the bivalence principle and the law of excluded middle.

Let us assume here that the law of excluded middle is a principle about existence, the claim, that is to say, made in the first clause of the existence principle that there is no alternative to existing or not existing.[14] I have been taking it for granted so far that the law so understood and the principle of bivalence, understood as the claim that every statement is either true or false, are no more than two forms of the same principle. The reason for this should be obvious enough. It is being assumed that to say that electrons exist, for example, is to say that the statement 'Electrons exist' is true, and to say they do not exist is to say that this statement is false.[15] The truth or falsity of statements, in other words, is being assumed to depend entirely on the existence or non-existence of things or states of affairs in the world. This is to adopt what has somewhat misleadingly been called the 'correspondence' sense of the word 'truth'.

The first point to be made here is that the word 'true' may be used in other ways. It may, in particular, be used so that it applies only to statements about which it is possible to gain knowledge. In which case we may indeed reject bivalence while retaining our principle about existence. Accepting the knowledge-dependence of truth, in that sense of 'truth', would be quite consistent with rejecting the knowledge-dependence of existence. All we need to do is keep well in mind that we are employing this 'epistemic' concept of truth, and cutting the tie between truth and existence through the adoption of a different terminology. Those who wish to detach metaphysical questions about existence from semantical questions about truth often do not make it sufficiently clear that this is what they are doing.[16] For our purposes, we may ignore semantical questions about bivalence when truth is understood in this way. Our concern is only with whether there is a tie between knowledge and existence, in our elementary sense of 'existence'. For this concern any issue about truth in this epistemic sense becomes irrelevant, and nothing at all is said about filling the gap of the unknowable.

However, even when 'truth' is used in the 'correspondence' way which makes it dependent on existence, the bivalence principle can be understood in a way which detaches it from our concerns. When it is expressed as 'every statement is either true or false', two things are being said which are detachable from one another. One is that there are only two truth-values, true and false. The other is that every statement has a truth-value. To reject the first of these is to claim there are alternative truth-values. To reject the second is to claim that some statements have no truth-value at all, that there are, as it is sometimes put, 'truth-value gaps'.

The distinction between the theory that there are alternative truth-values and the theory that there are truth-value gaps may seem unimportant to some, their concern being only with the fact that there is on both such theories a third class of statements to be coped with over and above the true and the false. But for anyone concerned with implications about existence, the distinction is crucial. To have any implications about existence the bivalence principle must be taken in the form in which it denies that there are alternative truth-values. Such alternative truth-values are required to correspond with alternative 'existence values', that is, with alternatives to existing or not existing. Merely to allow that there are statements without truth-value even if they are said to be meaningful, can carry no commitment about existence whatsoever, not even a commitment to some alternative to existing or not existing. If there is no truth-value, there is no existence value either. And there is no slot into which we can fit a mere absence of existence-value.[17]

The effect that this has on the connection of the theory of the knowledge-dependence of existence with the rejection of bivalence is not far to seek. Let us agree that if existence is dependent on the possibility of knowledge, statements purporting to be about what lies outside the limits of knowledge are neither true nor false. To that extent the rejection of bivalence follows. But there is never any suggestion that such statements have alternative truth-values. All that one can say is that being outside the limits they have no truth-value at all. The theory, in short, allows only for truth-value gaps. And what corresponds to a truth-value gap is not any alternative to existing and not existing, but what we might call an 'existence-value gap'. From certain statements having no truth-values, nothing whatsoever follows about existence or non-existence or any alternatives to these. The law of excluded middle, considered as a principle about existence, may possibly be challenged by its being said that there is more to the world than simply existing or not-existing, but it cannot be challenged by envisaging situations in which the whole question about existence does not arise.

It would seem then that anti-realists should stick to the issue about knowledge-dependent existence and not associate it with a challenge to the law of excluded middle. Their concern is with clause 3 of our existence principle, not with clause 1. The challenge to clause 1 on other grounds is something that we will have to consider in due course.

What I have mentioned here, it should be noted, is only one alleged special kind of existence—the dependent existence of objects. As opposed to this, minds may be said to have an independent existence, and it may be thought that independent existence should be treated as another kind of existence. But this is not necessary, and is probably not intended by any idealist or verificationist. It is true that, where there is a dichotomous division of a genus into those things which have a certain property and those which do not, it is commonly assumed that those which do not have it have some opposed property. If we divide elephants into Indian and non-Indian, we will expect the non-Indian to have properties which make them non-Indian, such as the opposed properties of African elephants. But in this case, it would seem, independence is not to be considered as a property of existence at all, but merely the lack of one. So long as the dependence is one-sided therefore, believers in dependent existence must admit that there is always one sphere in which the idea of there being kinds of existence does not arise and that is the sphere of those things on which dependent existents depend. The breach of the existence principle thus remains partial, but it is a breach of it nevertheless.

DEGREES OF EXISTENCE

Admission of kinds of existence is, as I have pointed out, closely connected with admission of degrees of it. So it is not surprising that the idea of knowledge-dependent existence may be seen as involving us not only in a special kind of existence for objects but also in there being degrees of it.

There is in the first place a decided inclination among idealists, if not among verificationists, to envisage knowledge-dependent objects as having something less than the 'full-blooded' existence of the knowers of them. On the reason for this, I will not bother to speculate. It certainly has nothing to do with an analogy with ordinary causal dependence. The existence of human beings is in the causal way wholly dependent on oxygen, yet this breeds no inclination to think of them as less real.

What is more important, however, is that knowledge-dependence may involve those committed to it in accepting degrees of existence in a way not normally recognised. For knowledge itself, as we move away from 'knowledge in the strict sense', is to be thought of as varying in degree in accordance with the degree of cogency in the evidence for a belief. It follows that those committed to knowledge-dependence have to accept degrees of existence in the objects of their beliefs to correspond with these degrees of cogency. If they do not admit this, they will be in the unhappy position of having to fix a quite arbitrary degree of justification for their beliefs as required for conferring truth on them—or at least, truth-value—not allowing as true—or as either-true-or-false—anything which falls below it. Since there is no reason for fixing such a limit they seem to be involved in a hierarchy of degrees of existence, whether they have envisaged this or not.

Berkeley, it is true, was able to avoid this conflict with clause 2 of our principle, because he was concerned with immediate objects which he took to be incorrigibly known. And again, Dummett's starting-point being intuitionism in mathematics, he was thinking in terms of the all-or-nothing concept of proof. But once you start dealing in degrees of evidence in the gaining of empirical knowledge, as in fact Dummett himself has been prepared to do, you run into this additional conflict with the existence principle.

THE PATH TO RELATIVISM

It is very easy to move from the idea of dependent existence to the idea of relative existence thereby involving oneself in yet a further breach of the existence principle. In fact in describing the dependent existence of objects of knowledge as existence *for* a knower, we have already included one element which goes into relative existence. Dependent existence, in other words, is a kind of existence which cannot be described without referring to its context, the context of that on which it depends. To that extent that which possesses it may be said to exist relative to that context.

At this point, however, it is important to make a distinction between a context that is invariant, and a context that varies in one way or another—from person to person, or theory to theory, or culture to culture. Idealists and positivists have been inclined to assume that the objects of knowledge all have their one kind of existence in so far as they are known by someone or other, or could be known by someone or other no matter who. Thus, for Berkeley, it was good enough to ensure the existence of an idea that it be in someone's mind. He did not say that my ideas existed only for me and yours only for you. And he backed up this point by saying that after all they all existed eternally in the mind of God, on whose knowledge they depended.

This assumption of an invariant context for knowledge-dependent existents was presented in a particularly conspicuous form by Kant. Kant of course was, as he himself insisted, no Berkeleyan idealist; for one thing he took it for granted that there were simply existing, though unknown, objects, as well as simply existing subjects of knowledge. For this reason he had to make quite explicit the distinction between this simple existence and the knowledge-dependent existence of the known world. To the transcendental reality of things in themselves he opposed the empirical reality of appearances, thus involving himself in the idea that things as known were things which possessed a special kind of reality. This was something to which he was committed as much as was any idealist. But in addition to this he developed in a much more explicit form the idea that this world of empirical reality was one world, dependent on an invariant context. For him it was no longer a collection of Berkeleyan ideas, it was rather a single world organised within the framework of a categorial scheme, common to all beings capable of judgment. Moreover, according to his transcendental argument, any such being had to adopt this scheme if it was to experience a world at all, so that there was a necessary guarantee of invariance of context for all knowledge-dependent existents.

Once, however, the tie is broken between known objects and an independently existing world, it may well come to be asked whether one can rely on invariance of context to ensure that there is a single knowledge-dependent world with only one kind of existence. The necessity of Kant's scheme may be challenged, for may there not be various different schemes, each providing us with its own world? People differ greatly from each other according to their culture, to the scientific paradigm in which their thinking about the world is embedded, or even to their particular theories. At the extreme limit, individuals all differ from each other, so that it might be supposed that, once Berkeley's divine guarantee was lifted, he could well admit the ideas of each individual existed only for that individual. So to varying degrees one might be tempted to break up the dependent existence of objects into a plurality of kinds each relative to its own particular context of knowledge.

The theory so arrived at may appropriately be called 'relativism'. It is not, it should be noted, a metaphysical theory, a theory, that is to say, about what exists or about which of our beliefs about the world is true. It is rather what one might call a meta-metaphysical theory to the effect that there are different kinds of existence belonging to different worlds, and therefore different kinds of truth. Each such world is knowledge-dependent, but they differ according to whether they depend on the knowledge of different individuals or groups, operating with different rules for the justification of their beliefs.

The word 'relativism' has of course been used in various more innocuous ways which do not bring the theory it refers to into conflict, as this one does, with the existence principle. But I call this the appropriate one, because it is that to which relativists are driven if they are to say anything new and interesting. We may trace the path from the innocuous and acceptable to the interesting but unacceptable in the following way.

We start with a 'factual relativism' which consists merely in pointing to the fact that people in different communities, or in different historical periods, have beliefs or codes of conduct which are incompatible with one another. Relativism in this sense is concerned with social facts which are hardly in dispute. It says nothing about whether these beliefs and codes are justified, and nothing about the truth of the beliefs. It leaves us with the possibility of asking the natural question which of two incompatible beliefs is justified, on the assumption that it could not possibly be both.

At this point, however, we may well move on from such factual relativism to scepticism. Given great differences between beliefs and codes we may be troubled about how we can ever tell which are justified. We are here put into the position of having to say that regrettably we will never know what to believe about the world or what we ought to do. This is an undesirable position to be in and we may well cast around for any way in which it can be avoided. This is the point at which a new and interesting way presents itself. This is to say that there is variation not only in actual beliefs and codes, but also in the standards which determine which are justified. Where there are incompatibilities, instead of saying with the sceptic that neither party is justified, we may say that both are justified, each using its own standards. This is the move which gives rise to relativistic theories not merely of beliefs but of knowledge and not merely of actual practices but of moral standards. It is the move to epistemological and moral relativism.

Moral relativism of this kind is undoubtedly more plausible than relativism about knowledge, and the reason for this is not far to seek. Moral standards are concerned with actions and not with the discovery of truth. The special difficulty which arises in the case of knowledge is the connection between the justification of beliefs and their truth. To justify a belief simply is to ensure that it is true, that things go on in the world in the way that we believe they do. How then can two beliefs which are incompatible both be true, even if we suppose that both are justified in terms of their standards?

It is at this point that we are driven to take the final step and move on from relativism about knowledge to relativism about existence. The two beliefs remain incompatible only so long as they are beliefs about the same world. If they are about different worlds with different kinds of existence, the incompatibility vanishes. The beliefs which are justified by your standards will be true of your world and those which are justified by my standards will be true of mine. The connection between justification and truth is preserved, but at the expense of the proliferation of different kinds of knowledge-dependent existence. This is the relativism which is in conflict with the existence principle.

SCEPTICISM AND KNOWLEDGE-DEPENDENCE

Given these widespread consequences of admitting knowledge-dependent existence, we may well ask what reasons anyone could have for admitting it. It is after all a natural assumption of ordinary life that what exists is one question and whether we can gain knowledge of it is quite another. The discovery of new facts about the world, and the confirming of new theories about it, seem to imply that there is something to be discovered. And this would still apply even if the discovery were shown to be in certain cases impossible. It is a sad thing, we may say, but there could well be some things about which we could never know anything at all.

It is true that this natural attitude may begin to weaken if the extent of our ignorance is alleged to be very large. We may accept quite easily, for example, that we may never know of what exists beyond the horizon of the universe. But if someone were to produce a good argument that we could never know anything at all about the world around us, we would begin to feel discomfort. As I have said when outlining the route to relativism, we find a wholesale scepticism an undesirable position to be in. The need to avoid scepticism, we may now say, is what drives people not only into accepting relativistic knowledge-dependent existence but into accepting knowledge-dependent existence of any kind.

It should be noted that I have here presented the move from scepticism to the acceptance of knowledge-dependent existence as based not on any argument but on the need to avoid discomfort. If someone does present you with a good argument for scepticism, you may take this no further but accept it, at least as a theoretical conclusion even though this is an unpleasant thing to do. To abandon the existence principle is to take a very large step further. This is a step which would only be justified if there were no objections to abandoning the principle to be weighed against the discomfort of remaining a sceptic. I suspect that the move has only been made because the central status of the principle and the difficulties of rejecting it have never seriously been considered.

In view of this it is not surprising to find that the standard arguments for knowledge-dependent existence have been arguments for scepticism and no more. Such arguments are not new. Since the time of Descartes, one predominant form they have taken is that of envisaging all of us as surrounded by a veil of our own experiences and our own thoughts, a barrier which we can never penetrate to gain knowledge of the world. Descartes dealt with this alleged predicament by appeal to divine guarantee, Locke dealt with it by ridicule, Hume by advising us to forget about it. It was Berkeley who made the first move from our alleged lack of knowledge of an independent world to its non-existence, and its replacement by a knowledge-dependent world. But apart from his attempt to justify this analytically by appeal to the meaning of the word 'exist', it was an anti-sceptical twist to his argument which remained unsupported. This traditional sceptical argument may not be a good one, but the point is that Berkeley did nothing to justify his way of avoiding it.

When we turn to relativism, we find that there is a similar reliance on sceptical arguments. Where different people hold incompatible beliefs, these arguments are to the effect that there is no way of telling which of them is true. What they show is that none of the parties are justified in holding their beliefs about the world. The extra twist which converts this conclusion into one to the effect that they are all justified in holding their beliefs, but about different incommensurable worlds, is not made the subject of a separate argument. To make this clear we need only look at the main arguments.

A basic argument of this kind is that from the underdetermination of our theories by the evidence. Whatever observations we make, it is always possible to explain them by different theories which are incompatible with one another. And in principle the conflict between such rival theories about the world will remain, however far we extend our observational tests. They should therefore be regarded as alternative frameworks in terms of which to view the world. In so far as they lack empirical control, there is no way of telling which is true.

This argument takes on a more radical form when it is applied not merely to conflicting theories but to conflicting rules for deciding between theories. Some say we should consult the oracle, others that we should seek confirmation in the greatest possible range of observations. Some would tip the balance in favour of the simplest theory, others prefer theories which are closer to the facts. In such cases, though there may be agreement about the evidence, there are still different criteria for rational belief and there is no way of telling which one to adopt.

While this kind of argument may induce scepticism about theories and methodological rules, there is another which extends the challenge to the observations on which they are allegedly based. This is the argument from the theory-ladenness of the observations. Observations, it is pointed out, are affected by the theories people hold, and therefore cannot be accepted as a neutral empirical basis for deciding between them. We can therefore never know not only which theories are true but which observation-based beliefs are true.

Furthermore, it is often added to this that there is not only an absence of neutral observations, but an absence of a neutral language in which to express them. Meaning-variance gets added to theory-ladenness to provide us with an even more radically sceptical argument. Not only do we not know which of our beliefs about the world are true, we do not even know whether we have the same beliefs, though expressed in the same words.

While these form one strand in relativist arguments, there is another quite distinct strand which derives from sociological considerations about beliefs. All the beliefs, it is claimed, that anyone has about the world are the product of the psychological and social conditions in which they come to be held. We must therefore rule out the possibility that they are ever caused by a consideration of the evidence for them. And since consideration of evidence is the only way of finding out which of them is true, we will never know whether any of them are true. As parts of ideologies, they may be regarded as social facts to be explained but they cannot be evaluated.

I hope it will now be clear that all these arguments, as commonly used by relativists, are on the face of it sceptical arguments. They are concerned to show that there is an absence of knowledge of the world, not that the world is knowledge-dependent, or that there are different knowledge-dependent worlds. I am not concerned here with whether they are sound arguments but only with whether they support the notion of a knowledge-dependent existence. And this they do not do. To do it, they must in addition involve themselves in a rejection of the existence principle. For this rejection they provide no argument. And since, as we have seen, there are powerful reasons for accepting the principle such argument would be needed.

I have here been presenting the case for a minimum realism, the point of which is to reject the idea that the existence of objects can ever be dependent on the knowledge of them. In doing this I have said nothing about what such independently existing objects might be. I have allowed that they might even be states of the knower's mind, and I have allowed for the sceptical conclusion that there are no such objects. To say more would be to enter into metaphysical discussion about what the world is like and epistemological discussion about the possibility of knowledge of it. These are not our present concerns, though something will need to be said about them in connection with our next problem.

NOTES

1 Note that the term 'idealism' is sometimes restricted to theories that include the first of these theses. See, e.g. Elliott Sober, 'Realism and Independence', in *Nous* (1982) p. 369.

2 See Jaegwon Kim, 'Noncausal Connections', in *Nous*, Vol. 8 (1974). Reprinted as Essay 2 in *Supervenience and Mind* (Cambridge, 1993).

3 See Ch. 8, pp. 139-140, also pp. 142-151.

4 See G.E. Moore, *Philosophical Studies* (London, 1922), Essay 1 'The Refutation of Idealism', pp. 2-5.

5 See Michael Devitt, *Realism and Truth* (Princeton, 1984) Ch. 2, for advocacy of adding such a clause to make a stronger realist thesis.

6 Berkeley, *Principles of Human Knowledge*, Sect. 3. Note that feeling an awkwardness, he added to this, 'Or that some other spirit actually does perceive it.'

7 Berkekey, *Principles of Human Knowledge*, Sect. 58.

8 On the flowers on the mountain, see A.J. Ayer, *Language, Truth and Logic* (2nd ed.) (London, 1946) p. 146.

9 See V.I. Lenin, *Materialism and Empirio-criticism* for extensive polemic on this point.

10 See Neil Tennant, *Anti-Realism and Logic* (Oxford, 1987) Ch. 11, pp. 111-118, for helpful suggestions on this question.

11 See e.g. Dummett, *Truth and Other Enigmas* (London, 1978), 'Preface', pp. xxx-xxxi.

12 Dummett, *op. cit.*, pp. 18-19. He here presents a picture of our investigations bringing into existence what was not there before, though it is not of our own making.

13 Dummett, *op. cit.*, p. 155.

14 For caution about this assumption see Ch. 6, pp. 92-94.

15 Elliott Sober in 'Realism and Independence', *Nous* (1982) makes this point (using the same illustration) in Footnote 2 (pp. 383-384) where he defends the idea of 'semantic ascent'.

16 Neil Tennant, *op. cit.*, is one of these. He claims to be a scientific realist but a semantic anti-realist. Only in Ch. 12 does it emerge that he takes the legitimate semantic notion of truth to be derived from 'establishes-as-true'.

17 The 'non-existent objects' of Meinong might look like a candidate. But they are not intended to be the objects referred to in statements without truth-value. On non-existent objects, see Ch. 4, pp. 49-51.

CHAPTER FOUR

APPEARANCE AND INTENTIONALITY

In discussing realism and relativism, our concern has been with the objects of knowledge. What is known must of course exist, otherwise we would not be speaking of knowledge.[1] We have been concerned only with the alleged dependent character of this existence.

We do however have experiences the objects of which might plausibly be said not to exist at all, and this raises a quite different issue which could affect our existence principle. Of such objects we might distinguish two kinds. In the first place there are the appearances of things. How the world appears to those who experience it often differs greatly from how it is. Often what we perceive is not there, or is different in character from what we take it to be. Often what we believe is in fact not the case. In these cases of non-veridical perception and false belief, what are we to say about the existence of what is perceived and what is believed?

In the second place, there are the objects of such experiences as imagination and desire, what is imagined and what is desired are not supposed to be part of the existing world at all, so questions about their veridicality or truth do not arise. Are we then in any way to admit their existence?

Those who wish to uphold the existence principle clearly have to deal with the problem about the appearances of things and with the more general problem about the existence of all those objects of experience, which since the time of Brentano have been called intentional objects.

APPEARANCES

That there are things which appear to be other than they are is a familiar fact of ordinary life. Everyone knows of the person staring at them from beyond the mirror, of the lake in the distance which disappears as it is approached, of the blue dome of the sky. And though not familiar to us now, there are things such as the caloric fluid which appeared to scientists at one time to pass from one bit of matter to another but which, so we now believe, does not exist.

Since these are familiar facts, there are some who have claimed that we should simply accept them as such, and have denied that they create a problem. John Austin, in his lectures on Sense and Sensibilia was one who gave forceful expression to this view. 'What is wrong', he asked, 'what is even faintly surprising, in the idea of a stick's being straight but looking bent sometimes? Does anyone suppose that if something is straight, then it jolly well has to *look* straight at all times and in all circumstances? Obviously no one seriously supposes this. To what mess are we supposed to get into here, what is the difficulty?'[2]

It is true of course that in ordinary life we accept the distinction between appearance and reality as a perfectly familiar one. But this does not show that there is no difficulty. All it shows is that in ordinary life most of us have neither the time nor the patience to consider the difficulty. In fact it may be said that the terminology of appearances is an extremely neat device, embedded in ordinary language, for registering the distinction while ignoring the difficulty. The distinction needs to be registered and we would no doubt not have survived if we did not recognise it. But it remains the case that, on the face it, the acceptance of it conflicts with another basic assumption of ordinary life. It is our good fortune that we can get along all right without resolving the conflict. But it is the fact of this conflict which constitutes the difficulty.

The other basic assumption with which the appearance-reality distinction conflicts is of course the existence principle. This principle requires that we raise the question whether appearances exist and demands an unambiguous answer. Yet in distinguishing the apparent from the real, it seems that we are avoiding the giving of such an answer. We do not want to say either that the apparent exists or that it does not. Does this then mean that in accepting the distinction, we have to abandon the existence principle.

At this point, we need to remind ourselves that the words 'appear' and 'appearance' may be used in different ways in different contexts. We need to do this in particular because there is one use in connection with which the difficulty we are discussing does not arise. This is the use in which we speak, for example, of a ship appearing on the horizon, or list the characters in a play in the order of their appearance. When 'appear' is used in this way, we are concerned to point out a relation between two straightforwardly existing things—the thing which appears and the person to whom (or before whom) it appears. No room is left for a distinction between the thing and its appearance. When we speak of a ship appearing on the horizon, it does not make sense to suggest that it appears as anything but itself; it simply appears.

When people are asked, however, *how* the ship appears to them, it is quite different. The question invites an answer like 'It appears small to me', or even 'It appears to me to be a large white whale'. We have here a use of 'appear' where it is no longer a question of something appearing to somebody, but of something appearing to somebody *as* so-&-so. The formula is not 'T appears to P', but 'T appears to P as A'. A third term has come into the formula. The thing-as-it-appears is no longer to be identified with the thing, and the question is forced upon us how we are going to fit this third item into our account of the world.

The problem so presented might be thought to be only about perceptual illusions, in which things recognised to exist appear to have characteristics which they do not have. But it would be a mistake to limit the problem in this way. There is after all no firm line to be drawn between such cases and cases of hallucination in which things appear to exist but do not. And in view of the well-known fact of the theory-ladenness of perceptions, there is no firm line to be drawn either between non-veridical perceptual appearance and false belief. The problem of the third item applies as much to the staring figure beyond the mirror and to caloric fluid as it does to the smallness of the ship.

The best way in which to assimilate these different cases is to include under existents not only things but states of affairs. Thus, there being a small ship on the horizon is a state of affairs which does not exist though it appears to do so. The Earth's being flat is a state of affairs which has appeared to some to exist, just as much as has the substance, caloric fluid. So we may, if we wish, treat all illusions as cases of hallucination, and all ordinary predicative propositions as existential ones. In all cases, we are faced with the problem of the third item—the thing or the situation which appears to someone to exist but which nevertheless may not exist.

Something's appearing to us to exist but not doing so should not, it must be noted, be simply identified with our failure to have knowledge of it. The appearance behind the mirror is, as everyone knows, quite compatible with a knowledge that there is no one there. There are many non-veridical perceptions which we automatically discount, but in doing this we do not eliminate them. It is true that in ordinary life we act on the initial assumption that how things appear is good evidence for how they are,[3] but this can easily be overruled by other evidence, such as looking behind the mirror. Non-veridical perceptions in short are not failures of knowledge unless we accept them at face value, and have no reason for believing them non-veridical.

Admittedly, in the case of beliefs the situation is different. We can have non-veridical perceptions which we know to be non-veridical, but we cannot have false beliefs which we know to be false. The appearance of caloric fluid disappears when the theory which supports it has been disproved. To that extent appearances, at the level of beliefs, are more closely tied to whether the beliefs are justified. But it remains the case that it is the possibility of beliefs being false which drives a wedge between how things appear and how they are, and this is quite different from their being known to be false, even though knowledge of their falsity eliminates them. In this respect, the analogy holds between false beliefs and non-veridical perceptions; in both cases it is the falsity or non-veridicality which causes the trouble, not whether we have reasons for or against accepting them.

The trouble caused by false beliefs and non-veridical perceptions can be presented in another way by pointing out that in these cases what is believed and what is perceived is incompatible with what is in fact the case. The distinction between appearance and reality is in fact forced on us by such incompatibilities. If a red ball appears grey, there is an incompatibility of colours. A body of translucent air and a blue dome cannot both exist in the same part of space. In the case of beliefs this incompatibility takes the form of logical inconsistency, such as that between the belief that the Earth is flat and the belief that it is not. If only there were some way of removing these incompatibilities, we could forget about the problem and use the word 'appear' in its first sense only.

In discussing the distinction between the world as it is and the world as it appears, I have let it be assumed that the appearances are the same for everybody. This is in fact often the case. There are standard illusions such as those I have mentioned of the person behind the mirror and the blue dome of the sky. But we must keep in mind that things may appear differently to different people. There are differences of perspective,[4] and there are illusions which are not standard. When it comes to beliefs the number of diverse false beliefs which are held by different people about any state of affairs can be very great. So there are not only incompatibilities between the apparent and the real. There are also incompatibilities between appearances themselves. In considering the problem this is something further which we must take into account.

So far we have been speaking in terms of specific perceptions and beliefs—of people behind mirrors, blue domes, caloric fluid and a flat Earth. It is in such cases that the problem of the unwanted third item most naturally arises. It is presented against a background in which we assume that the world is for the most part as we take it to be.

However the very same problem has often been thought of as emerging on a much grander metaphysical level. Material things in the world around us all appear to be coloured, but according to some they are not so in reality. They appear to us to be material, but according to some they are mind-dependent entities. There appear to be many things in the world but in reality it is said, it is a single whole. They appear to be in time but the only reality, it is said, is eternal. In older introductions to philosophy the job of the philosopher was sometimes described as that of penetrating behind the appearances to discover the real. And then of course the consequent question would arise about how to 'save the appearances'.

One who made great use of the appearance-reality distinction on this grand metaphysical level was Immanuel Kant. The empirical world in space and time was referred to consistently by him as the world of appearances (Erscheinungen). These of course were envisaged as the same for everyone; they were to be taken as a clear case of standard appearances. He did not think it possible, by the use of speculative reason, to penetrate behind these appearances to discover the reality. But he never doubted that there was one, for, as he remarked in a famous passage, 'otherwise we would be landed in the absurd conclusion that there can be appearance without anything which appears'.[5]

Another who made use of the distinction on an even grander level was F.H. Bradley, who in the late nineteenth century, following up themes from Hegel, wrote his treatise on Appearance and Reality, in which everything was branded as appearance except that single undivided whole which he called 'The Absolute'. It was appearance because it was self-contradictory and nothing self-contradictory could be real. In his case however he thought he had a method, derived from Hegel, of overcoming the contradictions and penetrating the veil.

SAVING THE APPEARANCES

The problem we are left with then, both at the mundane and the grand level, is what to do about appearances which, it would seem, are neither real nor nothing. In referring to them as a 'third item' I have tried to find a non-committal phrase, and it is important that we should be cautious. I have used such phrases as 'the world as it appears', and when I have introduced the word 'appearance', I could be understood as doing this on the assumption that it is no more than an abstract noun derived from the verb, so that describing the appearance of something is the same thing as describing how it appears. The question is left open here whether we can be pressured into using 'appearance' as a concrete noun, in which case appearances could be treated as entities which could exist even if there were no realities of which they were the appearances. Such a 'reification' of appearances is something which we do not normally envisage, and it was certainly not envisaged by Kant when he said it was absurd for there to be appearances without anything which appears. So we should consider first what to do about the 'world as it appears', where this is not to be simply identified with 'the world', yet is in its own special way dependent on there being such a world.

Things-as-they-appear seem on the face of it to be something of a metaphysical monstrosity. We know what it is to describe things, mentioning their properties and their relations to other things. And we commonly assume that we could give a complete account of the world in terms of things (or let us say 'entities') their properties and relations. It is not at all clear how things-as-they-appear can be fitted into such an account of the world.

It is at this point that the abandonment of the existence principle can be seen as a very natural way out of a predicament. Why not, it may be said, simply distinguish between real existence and apparent existence as two kinds of existence. An ordinary description can be given of the really existing world without mentioning appearances, and phenomenological descriptions, one or many, can be given concerned solely with the world or worlds of appearances. The important thing would be not to mix them up, since they would be descriptions of different worlds, and being descriptions of different worlds would not be incompatible with one another.

This would of course be a challenge to clause 3 of the principle. If appearances were different for different people, it would also be a challenge to clause 4. It could, moreover, be converted into a challenge to clause 1, if it were said that what really existed was all that existed, and that there was no such thing as the world-as-it-appears. Being an appearance would then become a third alternative to existing or not existing, and to include it would be to reject the law of excluded middle.

This, in one or other of its versions, looks like an easy way to solve the problem. If we do not adopt it in ordinary life, it is because we do not concern ourselves with the problem. But once we start to think about the problem, we may easily be pushed into it. We may play around with it for comic effect, as in the rhyme –

> Yesterday, upon the stair
> I met a man who wasn't there.
> He wasn't there again today
> How I wish he'd go away.

But if asked what has gone wrong here, we may well say that the author has failed to distinguish between what really exists and what apparently exists, or between what exists and what is only an appearance.

Not only is this an easy solution, but one which has been adopted even by the philosophically reflective. It has been adopted most spectacularly by those philosophers who have made use of the appearance-reality distinction at the high metaphysical level. This may be illustrated by the cases, already mentioned, of Kant and Bradley both of whom embraced it in its first version.

Kant, as we know, identified his distinction between things as they are in themselves and things as they appear to us with the distinction between the transcendentally real and the empirically real. How to interpret this latter distinction has been a matter of dispute. But the most obvious way to interpret it is as a distinction between kinds of reality, one possessed by a world which was dependent on minds for its basic structure, the other by a world of independent and unknown things. Such worlds could not simply exist side by side since one was the appearance of the other. Nor could the appearances simply be forgotten about, as we are inclined to do in the case of ordinary illusions or hallucinations, since, on Kant's view, they included our whole causal system of things in space and time. A 'two worlds' doctrine, therefore, seemed to become the inevitable solution to the problem.

When we come to Bradley, the idea that there were degrees of reality was no longer a matter of interpretation; it was an explicit part of the doctrine. The adoption of the appearance-reality distinction in fact might be seen as a way of giving an authenticity to the doctrine which it might not otherwise have had. But that the two fitted neatly together there could be no doubt. If the one real thing was the absolute whole, what fell short of it would have to be condemned as appearance only. Yet such appearances had to be saved, and to grant them their own kind of lesser existence was a way of doing it.

All this is on a grand scale, far removed from the person behind the mirror and the caloric fluid. But the problem of saving the appearances is the same. In either case what we need to realise is that rejecting the existence principle if it were allowable would be an easy way of solving it. I have argued however that it is not allowable. Though it looks like an easy way out it cannot be sustained against criticism. So if we are to solve the problem of appearances at all it must be done in some way which does not conflict with our elementary concept of existence.

This is the point at which something must be said about those metaphysical theories put aside as not of our concern at the end of the previous chapter. There our sole concern was to point out that objects of knowledge had an existence which was not knowledge-dependent. How further to characterise them was not at issue. Now however that we have moved on from objects of knowledge to appearances and need to avoid granting to them an inferior kind of existence, we need to consider how this might be avoided.

There can only be two ways of doing this, both of which are to be found embedded in certain traditional and current theories about perception and belief. One is to say that appearances exist in the ordinary way, in addition to, or even in replacement of, those things of which they are the appearances. The other is to deny that they exist at all.

The first of these has been the most common. It follows the path I have mentioned of taking the noun 'appearance' as a concrete noun. It can then be the name of an entity which exists in the elementary sense. Given that there are such entities, there is the need to find a place for them in the world, and the obvious place is the minds of those for whom they are the appearances. Hence the 'ideas' of Descartes, Locke and the rest. Hence also, though more non-committedly about where to place them, the 'sense-data' and 'propositions' of modern representationalists and phenomenalists. Given such entities, however placed, there will be no problem about incompatibility, since the incompatible properties which bothered us will belong to different things.

This way of dealing with the problem has certain consequences which should be noticed. In the first place it no longer seems reasonable to restrict appearance to those cases in which things appear *other* than they are, that is to say, to cases of illusion and false belief. So far we have let it be assumed that when things appear as they are, this is the same as their simply appearing, in the first sense of 'appear', the relation being a two-termed one and no third item being involved. But once we set up appearances as entities, we lose any motive for discriminating between the two cases. What we have before us are the items we call appearances, and whether they are veridical or illusory, true or false, depends entirely on the relation which holds between them and the reality, on whether, as it is sometimes put, they represent this reality correctly or incorrectly.

In the second place it should be noticed that when appearances are entities existing in this way, they no longer need to be appearances of anything. What Kant considered absurd—that there could be appearance without anything which appears—is no longer absurd but a possibility to be envisaged, as has been realised well enough by the phenomenalists. What has happened is that the original notion of appearance has been analysed in terms of one entity representing another, and it is a contingent matter whether there is anything at all to be represented. To that extent one element in the original notion has been abandoned. We are still left of course with the notion of appearing in the first sense, because now the appearances appear before us in the same way that the ship was said to appear on the horizon.

In the third place it should be noticed that appearances, conceived of as entities, are something we can have knowledge of. Since their appearing before us is a simple two-term relation, such knowledge is naturally thought of as immediate and therefore incorrigible. Our mistakes about the realities are to be put down to our failure to realise that the appearances misrepresent the realities.

Once we think of appearances as objects of knowledge we may begin to wonder whether this reification of appearances does after all enable us to avoid a violation of the existence principle. For are we not now crediting them with that special sort of existence we criticised in the last chapter—a knowledge-dependent existence? Are we not envisaging them as a somewhat queer sort of entity which only exists when it is before the mind, though not created by it?

If we are to adopt this solution of our problem, we should certainly be careful not to fall into this trap. One suggestion for avoiding it might be to regard these entities as strictly mental, that is to say as parts of the mind rather than its objects. As parts they could be credited with simple existence along with the whole of which they are the parts. The trouble here, of course, would be that, on this view, they would be a special kind of parts which are not detachable from the whole, and therefore still open to the charge of having a dependent existence. The move from objects to parts does, however, open up a way of countering this. We might say that the dependence of the parts on wholes is no longer mysterious but merely a causal dependence of a kind that we find in all existents that are organically related. We admittedly do not create the appearances in the sense that we determine their character. This determination comes from things outside us or at least, in the case of hallucinations, from some bodily condition affecting the nervous system. But their organic relation with the mind as a whole could be regarded as another causal factor necessary to account for their existence. This would give us a harmless account of their non-detachability.

There is however another and simpler way of making the reification of appearances consistent with the existence principle. This is to retain their status as known objects, yet allow that they are detachable. As independent entities, they could, on this view, exist as appearances without appearing to anyone. They could still be thought of as third items representing other things, correctly or incorrectly. But there would be nothing to stop there being a vast array of them, mostly not objects of anyone's knowledge. And this suggests that, having treated them as entities, we might even elevate them into being the sole entities, comprising the world instead of representing it, the illusory and the false being merely those among them which did not fit into a simple and coherent system with the rest. Thus phenomenalism could be added to representationalism as a theory which copes with appearances by treating them as straightforwardly existing items.

These are familiar theories, about which much has been written. But their treatment of appearances as entities is something one would wish to avoid if one could. We should therefore draw attention to the other type of solution which consists in denying that appearances exist, either as third items in the world, or as substitute second items comprising the world. On this view appearing simply remains a two-term relation between things in the world and those to whom they appear. To take this view is not to ignore the problem but to attempt to meet it straight on. The trouble is that those who maintain it still have to account for illusions and false beliefs, that is, for cases where things appear other than they are. And given their view the difficulty of doing this seems at least as great as that which confronts those who make appearances into separate entities.

One way of doing it is to claim that what is presented as a third item is really nothing more than a breakdown in the relation of appearing that we have with the world. Where things appear to us, we have knowledge of them. The relation being a two-term one, such knowledge is immediate. The trouble is that, being immediate, it is thought to be incorrigible, so that there is no possibility of making a mistake when things appear to us. When what appears to us is thought to be the third item itself, an idea or sense-datum or proposition which we believe, this assumed connection between immediacy and incorrigibility looks plausible. How after all could we possibly make mistakes about the contents of our own minds? But if we abandon such internal third items, the idea that the possibility of error is present in any knowledge relation is one that anyone should admit. And cases of illusions and false beliefs are on this view no more than cases where such error occurs. Their incompatibility with what is actually the case is, as it were, dissolved into a fault in the relation between the mental attitude and the world.

If this 'direct realist' solution to the problem could be maintained it would, I am sure, be a great relief to all of us. One trouble of course lies in the fact I have mentioned that illusions persist when we know they are illusions. If we could assimilate illusion to false belief, this trouble might be avoided since false beliefs do not persist in the same way.[6] But we are left with our false beliefs. If I believe the false proposition that it has rained more today than yesterday, is this really nothing more than a failure in my relationship to the actual state of affairs? Do not I still believe something, just as I still see something behind the mirror? Am I not left with an incompatibility between what is believed or what is seen and what is in fact the case?

With such difficulties in mind, we have to realise that solving the problem within the limits set by the existence principle is not plain sailing. But those who attempt this are at least facing the problem, not evading it. We might perhaps be helped if we now broaden the topic by taking into account not only appearances but all intentional objects including the second kind I started by mentioning.

INTENTIONALITY

There is not only an issue, as I pointed out, about what we perceive and what we believe, but also about what we imagine, what we desire, and so on. In all such cases, there is an object of some mental attitude or other. The difference is that in the case of perception and belief (and of course also memory) there is some relation to a world in virtue of which we distinguish between the veridical and the non-veridical, the true and the false. Where this is absent, the problem appears in a somewhat new light.

The problem however still remains as to what we are to say about the existence of such objects. They can no longer be regarded as appearances of something else, but they nevertheless remain as something that has to be accounted for. They are not third items which have to be fitted into a picture of the world and ourselves, but they remain as second items which have to be fitted into a picture of ourselves.

Since the time of Brentano, as I have mentioned, such objects have come to be referred to as intentional objects, in a technical sense of 'intentional' derived from the scholastics, and in that way have been distinguished from the real objects existing in the world around us. They include all kinds of objects such as the golden mountain I imagine, Mr Pickwick as described by Dickens, the destruction of all nuclear weapons which many of us desire. Brentano said they had an intentional inexistence, but what he was committing himself to in the use of this phrase was far from clear.

It will be obvious however that talk of such objects invites some breach of the existence principle in the same way as does talk about appearances. We have attitudes towards them, and this suggests some kind of a relation we have to them. Both terms of a relation must exist. Yet these objects form no part of what we would normally consider as the existing world. Why not then suppose that they have their own kind of existence in other worlds, worlds of imagination, of fiction, of desire? In so far as we envisage such worlds as not existing apart from our awareness of them, they may come to be thought of also as knowledge-dependent worlds, and this adds to the need to attribute to them a special kind of existence. Furthermore so far as these objects exist in different worlds each relative to its own 'universe of discourse' they may come to be thought of as having a merely relative existence, their kind of existence being dependent on their context. So as in the case of appearances we may envisage here a breach not only of clause 3 but also of clause 4 of the existence principle. There is still of course, as in the case of appearances, the alternative of rejecting the law of excluded middle, thereby breaching clause 1 instead.

As in the case of appearances then we need to consider what can be done about the problem without breaching any clause of the existence principle. Are we simply to grant existence or deny existence to intentional objects?

When we include such objects as those of imagination and desire, we can see at once that there is a special difficulty about simply denying their existence. This is because these objects do not have a relation to the world of the same kind as that of appearances to what appears. My image of a golden mountain has nothing to represent and is neither veridical nor non-veridical. A statement of fiction is neither true nor false. We cannot therefore eliminate the intentional object in favour of a direct relation to the world in the way that some have proposed to eliminate appearances in favour of a direct relation of appearing.

It is true that some attempts have been made to do this. It has been suggested for example that in cases such as that of imagination we are in fact experiencing selections from the world, though we put them together in a new arrangement.[7] The trouble with this is of course that when we speak of 'the world' we include in this its actual arrangement, and this is something we do not experience in imagination. The world admittedly includes gold and mountains, but it does not include golden mountains, and the attempt to regard them as derivatively part of the world seems doomed to failure.

Does this mean then that we are driven back to the other alternative, that of treating intentional objects as actual entities and finding a place for them in the actual world? Before committing ourselves to this, we must consider another and more spectacular way in which their existence has been denied.

NON-EXISTENT OBJECTS

It was Meinong who in developing Brentano's idea took the course of insisting that being an object was one thing and existing was another. He admitted intentional objects as objects, insisting moreover that they were independent of the various attitudes we have towards them, but said simply that they did not exist. This enabled him to avoid in one sweep any breach of the first four clauses of the existence principle.

The idea was that of a realm of objects which were 'beyond being and non-being' in the sense that they could have either, some existing but many not existing. Those that did not exist could still have properties; the golden mountain after all has the property of being golden. Sosein, as Meinong expressed it, was independent of Sein.[8] And one way of expressing Meinong's idea is to claim that it is possible to have properties without existing.[9]

So while rejecting kinds, grades or the relativity of existence and retaining the law of excluded middle,[10] Meinong issued a direct challenge instead to what I have listed as clause 6 of the existence principle and by implication a challenge to clause 5. It is a challenge which has been taken up, supported, and elaborated upon in recent years, despite the opinion of many that it had been effectively countered by Frege and by Russell.[11] It also appears to be implied by the proposal of phenomenologists to study objects of consciousness after having excluded their existence.[12] We need therefore to consider seriously whether it provides us with a further alternative way of solving the problem of intentional objects, despite the fact that it continues to reject the existence principle in other of its clauses.

For many this detachment of being-an-object from existing has an air of paradox. How, we feel, can anything be an object without existing? Hence is not the theory simply equivalent to the self-contradictory assertion that non-existent objects exist?

The reply to this will of course be that we feel this way because we implicitly accept the existence principle, and it is this principle which is at issue. Furthermore one may appeal to the counter-intuition that when we imagine, tell a story about, or in general refer to, say Pegasus or Sherlock Holmes, we are in fact imagining, describing or referring to something which does not exist. Indeed the principal argument in favour of the view that there are non-existent objects has always been that we do constantly imagine and refer to things, while happily admitting their non-existence.

I have here introduced the notion of 'reference', and have bracketed it with the mental attitude of imagination. Similarly, when I was introducing clause 6 of the existence principle in Chapter I, I bracketed it with experience, which I there used in a provisional way for any mental attitude towards an object. Yet it should be noted that when we speak of 'reference' we introduce a linguistic twist to the issue about intentional objects. We refer to objects by using words, in particular proper names and descriptions. So the issue becomes one about what such words as, say, 'Pegasus', or 'Sherlock Holmes', or 'the king of France' refer to. The argument is that they obviously do refer to something though it is recognised that what they refer to does not exist.[13]

The theory that admits of objects which do not exist has in fact often been discussed in terms of the reference of phrases, rather than objects of mental attitudes. And it was its presentation in these terms which made possible the first major criticism of it. In his essay 'On Denoting', Bertrand Russell described Meinong's theory as one which 'regards any grammatically correct denoting phrase as standing for an object'.[14] It was against the theory so described that he claimed to show that denoting phrases, such as 'the present king of France', did not require such objects, but should rather be interpreted in terms involving assertions (or presuppositions, in Strawson's later version) about the instantiation of properties, or the values of variables.

Let us assume, as many have, that this interpretation is correct and that there is no need to assume any object, existent or non-existent, to which phrases like 'the king of France' refer. Unfortunately however this does not help to solve the more general problem about those objects which we think of, imagine, or desire. It may well be true that the use of the phrase 'the golden mountain I am thinking of' does not in itself involve us in assuming any object to which the phrase refers. But this does nothing to counter the claim that in this case I am thinking of an object and that in thinking of it, I am neither asserting nor presupposing that there exists anything which instantiates its combination of properties.[15]

We return then to the crucial evidence alleged to be in favour of certain objects of thought not existing and that is that we have a variety of mental attitudes towards what we on other grounds know to be non-existent. Our problem is how to challenge this evidence without presupposing the existence principle.

At this point one thing we can do is to re-introduce certain considerations about relations. We pointed out, in Chapter 2, that anyone envisaging things having different kinds of existence must have difficulty about there being any relations between them, whether of dependence or anything else. We can now point out that, a fortiori, there is a more extreme difficulty about existents such as mental attitudes being related to what has no existence at all. For any relation there must be at least two terms. And it is hard to see how there could be such a relation at all unless both these terms were entities of one kind or another. If it is said that in this case the relation is of a somewhat exceptional sort, in that one of its terms does not exist, one can only conclude that the word 'relation' is being used in a very unusual way, and that this shift in usage is being misleadingly represented as a distinction in kind. If my being afraid of a ghost is taken to be a relation between me and the ghost and it is added that the ghost does not exist, it is very easy for us to be left with the impression that there is something there, in a sort of a way, for me to be afraid of. Yet, if there is no ghost, it follows that there is in fact nothing there for me to be afraid of.

Considerations such as these suggest that while no one doubts that people fear ghosts and imagine golden mountains, these facts of life need to be interpreted in ways other than by retaining the ghosts and the golden mountains as independent objects and then denying their existence. And, as we know, one way of doing this is to regard them as existing items of some kind or other which constitute legitimate terms for the relations of being afraid of or desiring or whatever. We may call them 'representations' if we like, but we have to remember that in this case there is nothing for them to represent. On such a view they are indeed dependent entities, but the more we move away from appearances, even hallucinatory ones, to other intentional objects, the more obvious it becomes that the dependence is causal in character. We not only provide the necessary conditions of the representations of imagination and desire. We also determine their nature.

OTHER SOLUTIONS?

Are other solutions possible? Clearly we must stick to our dichotomy. About anything we can name we must say either it exists or it does not, and that applies to intentional objects as much as to anything else. The features attributed to them may however vary. Furthermore if the variation from what we started with becomes sufficiently great we may choose to cross the line and say they do not exist but only that something like them does. Where we draw the line is a matter of choice. But once it is drawn we may speak of the crossing of it as the reduction of intentionality. The limit is reached when we simply deny its existence without replacement. It is then that we speak of its elimination.[16]

The loosening up of the alternatives in this way is well illustrated in the discussion of this issue in recent years. The terminology has changed, the issue generally being presented as one not about the objects of experience but about the contents of propositional attitudes. But the problem is the same. It now becomes one of what to say about these contents.

There is one account which continues with the basic idea of regarding them as representations, existing like everything else. They are however characterised in a new way. Starting with beliefs and desires as basic attitudes, their contents are described, according to one dominant theory, as constituting a kind of internal language, the 'language of thought'. This 'mentalese' moreover may be thought of as embodied in the structure of the brain. Representations so characterised may seem far removed from those mental pictures of ghosts and golden mountains which many have thought of as representations, but they may nevertheless be viewed as the modern propositional descendents of Cartesian ideas.[17]

We pass on to a more reductive account of the contents of our mental states when they are said to consist not of objects like internal sentences but of the connection of these states with others in a special kind of network. Talk of what is believed and what is desired is to be understood, on this view, in terms of the part the beliefs and desires play in the mind (or the brain) as an intentional system, in which stimuli are suitably interpreted and associated with appropriate responses.[18]

Having come so far one might wonder whether one had not better stop talking about intentional objects at all. And in fact this is the final eliminative move which has been made. It is more radical than earlier attempts to eliminate intentional objects in favour of actual objects in the world around us. It consists rather in denying the existence of intentional states altogether. On this view the theory of the mind which involves us in attributing to it beliefs and desires and their objects is not only just a theory but a false theory. It gets us by in ordinary life just as does the theory that the sun rises and sets over a stable earth. But as psychological theory such folk psychology needs replacing by the more adequate theories to be developed by a cognitive science which has a sound neuro-physiological basis.[19]

It is not our business to enter into a discussion of these theories. But a final remark needs to be made on the legitimacy of saving the appearances by eliminating them in the last way mentioned. False theories may be eliminated, in the sense that they may be shown to be false. But, even if false, our commonsense theories about beliefs and desires continue to be held by many people, just as common-sense theories about the movement of the sun were once held. They remain in short as intentional objects or propositional contents which have to be accounted for. False beliefs, like any other intentional objects, cannot be simply eliminated any more than they can be relegated to some shadow world of half-reality or to some world of non-existent objects. We seem then to be in the paradoxical situation of having to find a place for them or for some replacement of them in the existing world, whether this be in the form of representations or elements in an intentional system or anything else.

Apart from this warning shot about the intractability of false beliefs, I have not been concerned to argue for or against any of these theories on their merits. My point has been to consider them as possible ways of solving the problem about appearances and about intentionality generally, within the confines of the existence principle. Solutions have been offered. The problem of deciding which to accept has to be faced and not evaded.

NOTES

1 Knowledge of the truth of propositions has of course to be included here. On the connection between truth and existence, see Ch. 3, pp. 39-40 above and Ch. 6, pp. 92-94.

2 J.L Austin, *Sense & Sensibilia* (Oxford, 1962) p. 29.

3 This is the basis of what Frank Jackson calls the epistemic use of 'looks'. See F. Jackson, *Perception* (Cambridge, 1977) Ch. 2, Sect. 1.

4 For consideration of perspectives see Ch. 5.

5 Kant, *Critique of Pure Reason*. Preface to the Second Edition, BXXVII Kemp-Smith translation.

6 See D.M. Armstrong, *Perception and the Physical World* (London, 1961) Ch. 7, for an attempt to do this.

7 Samuel Alexander, *Space, Time and Deity*, Book 3, Ch. 8, contains an extended attempt to maintain a 'direct realist' account of all our experiences in this way. See esp. the section on Constructive Imagination.

8 Alexius Meinong, *The Theory of Objects*, reprinted in R.M. Chisholm (ed.), *Realism and the Background of Phenomenology* (Free Press, 1960), p. 82.

9 It is discussed in this form by Timothy Williamson, 'Equivocation and Existence', in *Aristotelian Society Proceedings*, 1987-8.

10 With the proviso that he preserved a kind of existence for universals which he called 'subsistence'.

11 See e.g. R.M. Chisholm, 'Beyond Being and Non-Being' (reprinted in *New Readings in Philosophical Analysis*, ed. Feigl, Sellers and Lehrer (Appleton-Century-Crofts, 1972)), R & V Routley, 'Rehabilitating Meinong's Theory of Objects' (*Revue Internationale de Philosophie*, Vol. 27 (1973)), Terence Parsons, *Non-Existent Objects* (Yale U.P., 1980).

12 As is often the case, there is some uncertainty here. Husserl e.g. describes the domain of objects of consciousness, at one place, as a 'region of being which is of essential necessity quite unique' (*Ideas*, Book I, trans. Kersten p. 65 (Nijhoff, 1982)). Has it a kind of existence after all?

13 See, e.g., Nicholas Wolterstorff, 'Referring and Existing', *Philosophical Quarterly* (1961), for an example of the argument being presented in this way.

14 Bertrand Russell, 'On Denoting', reprinted in *Logic and Knowledge*, ed. R.C. Mars (Allen & Unwin, London, 1956) p. 45.

15 See R.M. Chisholm, 'Homeless Objects', in *Revue Internationale de Philosophie*, Vol. 27 (1973) p. 208, for effective presentation of this point.

16 On reduction and elimination see J. Schwartz, 'Reduction Elimination and the Mental', in *Philosophy of Science* (1991, pp. 203-220), and discussion of it by Colin Cheyne, 'Reduction Elimination and Firewalking', in the same journal (1993, pp. 349-356).

17 See J. Fodor, *Representations* (Harvester, 1981) esp. Ch. 7 on 'Propositional Attitudes', for an exposition and defense of a view of this kind.

18 Paul Churchland, *Matter and Consciousness* (Cambridge, Mass., 1984) Chs 3 & 4, has an account of the use of a network theory of meaning to solve the intentionality problem. See also Daniel Dennett, *Content and Consciousness* (Routledge, 1969), Chs 3 & 4, for the original development of this idea.

19 See Paul Churchland, 'Eliminative Materialism and the Propositional Attitudes', in *Journal of Philosophy*, Vol. LXXVIII 1981, pp. 67-90, for a defense of this 'eliminative' as distinct from 'reductive' solution of the problem.

CHAPTER FIVE

PERSPECTIVES AND PARTIAL TRUTHS

In discussing appearances, we have noted different ways in which we might try to deny their existence, thereby avoiding danger of conflict with the existence principle. There were in fact three such ways—the 'direct realist' way, the 'non-existent-object' way, and the way which rejects the whole of folk psychology as false. All of these, as we saw, faced serious difficulties. In dealing with the first of them, however, we did no more than mention one idea which might on consideration enable us to reinstate it as an acceptable solution to the problem. This is the idea of a perspective.

In considering this idea we should start by pointing out how extremely common it is to refer to people's perceptions of the world around them, and also to their beliefs about it, as *views*. People who hold these views are said to adopt certain *standpoints*, or to take up certain *positions*. All of which can be summed up by saying that each regards the world from a certain perspective. Each has a view of the world—a Weltanschauung. Their views are different, but the world is the same.

Is it not the case then that the very same thing can after all appear differently to different people without it in any way appearing other than it is, or the appearances being incompatible with one another? All we need to do is put the difference down to a difference of point of view. This gives us a kind of a bridge between the two uses of 'appear'. We are concerned with two straightforwardly existing items, the world and ourselves, but the perspectival nature of the relation between them enables us to reconcile not only the world with its appearance, but also the singleness of the world with the multiplicity of its appearances. What seemed to be impossible would not, on this line of thought, be impossible after all.

It is not surprising, therefore, that the idea of a multiplicity of perspectives has been a very attractive one. It is however a very dangerous one. It is not only that the appeal to it fails to solve the problem about appearances. It is also that once taken beyond its own limited field of literal application, it can become a very misleading metaphor, which serves to sanction, and lend plausibility to, further breaches of the existence principle. We should therefore pay some attention to its origins and to the various ways in which it has been misused.

PERSPECTIVES AND SPACE

The origin of the idea of a perspective lies obviously enough in our visual perception of things. It is in visual perception that things appear differently to different people according to their point of view, depending on their angle of vision and their distance from the thing. And the reason for this special perspectival character of vision lies in the nature of space.

Space is a continuum which allows of a certain kind of relational structure for things within it. Everything in space has a position of its own, distinct from the positions of other things. Given that position, a thing has an environment consisting of all other things in space, and to each of these it has a different spatial relation—near or distant, north or south, and so on. We are assuming here of course that any two things share a common environment; we are thinking, that is to say, of a public space. So if there are two people in a room, each is related by different spatial relations to other things in the common environment, such as, say, a chair or a table.

All this, it should be noticed, does not require that the space be three dimensional. A two-dimensional surface will do. It is this which makes it possible to represent the spatial relations between things in two-dimensional diagrams. Graphical representations, with co-ordinates and points of origin, make use of this point.

It may be said that the space I have here described is Newtonian space and that account has not been taken of the new conception of space embodied in the theory of relativity. But this is in fact not so. The space that has been described is the space of common sense. It is this which forms the basis of our talk of views, standpoints and aspects. What further theoretical account we are to give of it, Newtonian or relativistic, is therefore not to the point.

Given this character of space it is a small step to convert positions within it into points of view. All we need to do is to put people into the positions and endow them with eyesight so that they can look at things in their environment. From their different points of view each can look at the same things, but for each of them their spatial relations to these things are different so that the things look different. It is true that no single person can adopt two different standpoints at the same time. But there is no problem about the same things being viewed from different standpoints by different people or by the same person at different times.

That it is the character of space and not of vision which is responsible for this becomes clear if we consider that the same situation holds in the case of movement in space. Two people may not only look at a thing from different points of view, they may also approach it along different paths. Two people may not only observe one and the same mountain from different aspects; they may climb one and the same mountain by different routes. Admittedly neither can climb it by both routes at the same time. But there is no problem in one taking one path and the other taking another to reach the same destination. Hence it is not surprising that the metaphor of movement is often used in the same way as that of perspective vision, as when we speak of there being different approaches to a problem, or someone following a path which leads to a conclusion. When we speak of perspectives therefore we should keep it in mind that in certain contexts approaches can do the same job.

Because of these facts about space, we must agree that there is one context in which it is entirely appropriate to give an account of appearances in terms of perspectives, and that is the context of visual perception. In so far as the way in which we see things depends on relative positions in space, no awkward questions need arise about third items, or of things appearing other than they are. Nor will there be any incompatibility between the different appearances of the same thing.

A MISLEADING METAPHOR

Trouble arises immediately, however, when an attempt is made to extend the notion of perspectival vision to other kinds of case in which things appear differently to different people. When this is done talk of perspectives loses its special function, and becomes a metaphor. When metaphors are accepted as metaphors, they may be harmless. But they may also be very misleading.

All metaphors are based on some analogy, and of course there is an analogy between visual perceptions involving perspective, and all other perceptions and beliefs. In all cases some claim is being made about the nature of the world. Furthermore, the claim is in all cases being made by someone who is part of the world. Just as all perspective views are from within a spatial environment, so all perceptions and beliefs are those of people living in a specific physical, psychological and social environment. In the one case as in the other, differences of environment affect the claims made, and in each case there is no 'view from nowhere'. This is an analogy which has had considerable influence in the expansion of the 'standpoint' terminology. Views from positions have been generalised into views from situations. And since everyone is in one situation or another, the terminology can then be applied to every perception and belief.[1]

It remains the case, however, that when all perceptions and beliefs are referred to as views of the world from given positions, an important dissimilarity is being passed over, and it is that which makes the metaphor misleading. It is a dissimilarity which is quite unaffected by the fact that positions in space are one kind of situation and that all views of the world are views from 'within'.

The dissimilarity emerges as soon as we consider that seeing things in perspective leaves no room for seeing them as other than they are. From each perspective we see the same thing with the same spatial qualities, though some of these may be seen from some angles and distances, and some from others. Whereas in all other cases we have to admit the possibility of illusion and false belief, of things and states of affairs, that is to say, appearing as other than they are. Thus the person behind the mirror is not ourselves seen in a particular perspective; it is a person seen by us as occupying a position which is not so occupied. The dome of the sky is not a body of translucent gas seen from a given distance; it is to all appearances a solid dome. The man on the stair is not there, there is only a body of empty air, but it is seen as a man.[2]

The case of the ship seen as small may seem to provide a counter-example to this separation of perspective vision from illusion. For is not the smallness of what is seen just a matter of the distance from which we see it? The answer is that we need here to make a distinction which we are inclined to ignore when we use comparative terms like 'small'. Where the ship's appearing small is taken as a case of non-veridical perception, 'small' is taken to mean 'smaller than its actual size'. The ship has an actual size—say, 50 metres long—and if it appears smaller than this, it appears other than it is. But if this were merely a question of perspective—of seeing the ship from near or from far off—the whole question of comparative size would not arise. The ship would appear as it is, though differently to someone seeing it from 100 metres distant and to someone seeing it from 2 kilometres. It is true that someone might be deceived by the distance into taking it to be smaller than its actual size. But this error would be due not to the ship's being seen in perspective, but to its being seen in an incorrect perspective.

If the idea of perspective cannot cope with illusions it can cope even less with false beliefs. When it comes to beliefs there is not even a special context, such as vision, in which perspectives have their place. Believers in a flat Earth cannot, even metaphorically, be said to be seeing the globe from a point of view. Caloric fluid is not heat transmission by vibration of particles viewed from a different standpoint. Yet strangely enough, it is in connection with people's beliefs about the world that the metaphor of views and positions has been made great use of.

It is when things appear other than they are that the appearances, as I have said, become incompatible with the things and with one another. So the failure of the perspective analogy to cope with any such cases may be represented as its failure to cope with such incompatibilities. The special point about perspective differences is after all that they are reconcilable with one another and with the nature of the perceived object. To regard other appearances as no more than perspective views is thus simply to ignore these incompatibilities.

In the case of perceptual illusion, as I have pointed out,[3] these incompatibilities are not mere logical inconsistencies but depend on such principles of order as that nothing can at the same time have two different colours and that two things cannot at the same time occupy the same space. It is when we come to believe that the incompatibilities become directly logical, as when someone says that caloric fluid passes between substances and someone says it does not but that heat is transmitted in some other way. When such inconsistent beliefs are described as different views about the transmission of heat, the impression is given that the beliefs can somehow be reconciled with one another in the same way as can different views of objects in space. This is where the analogy becomes seriously misleading.

The idea that by talking of perspectives we can reconcile contradictory beliefs is one, moreover, which has been made use of not only on the mundane but also on the grand metaphysical level. We may refer here again to Immanuel Kant. I have mentioned that there can be some dispute about how his global distinction between appearance and reality is to be interpreted. And there are hints that he would have liked his world of appearances to be not a 'third item' but simply the real world of things in themselves as viewed 'from the human standpoint'.[4]

The analogy I am here criticising, it should be noted, is between visual positions and beliefs, not between the people who are in those positions and who hold these beliefs. It is as well that we should leave it that way, because it enables us to avoid a further failure of the analogy which might distract us. As I have mentioned, one person can only be in one position in space at any given time; this is the principle of the alibi. Whereas in the case of beliefs, we know very well that the corresponding principle does not hold, and that the one person can hold two inconsistent beliefs at the same time. This might provide a further reason for being suspicious of the analogy. But the important point for us is the inconsistency of the beliefs, whoever holds them, and its failure to fit with the reconcilability of perspective views, whoever the viewer.

It may be said that the disanalogy I am here pointing to should not be taken too seriously. We do not, after all, find it odd when people are said to hold conflicting views about a given situation or when one says of the other's view that it is incorrect. As a result of the positive analogy I have mentioned, the word 'view' has come to be more or less a synonym for 'belief' and its metaphorical origin has passed into the background. In so far as this has occurred, we have admittedly no need to worry. But there remain many contexts in which the metaphor retains its influence. If you are said to be looking at things from one perspective and your opponent from another, a strong impression is created that the admitted conflict between you can in some way be overcome.

PERSPECTIVES AND GRADES OF EXISTENCE—PARTIAL TRUTHS

So far nothing has been said about the perspective metaphor sanctioning any breach of the existence principle. What it has been shown to sanction is the idea that one and the same thing can have incompatible properties. At first sight this suggests no more than a rejection of the logical law of non-contradiction and of such principles as those of exclusiveness of colours and of space occupancy. Neither this logical law nor these principles are requirements of the elementary concept of existence. As far as this concept is concerned two things might exist in the same place, and it might be the case that something both exists and does not. All that we have to exclude is any alternative to its existing or not existing. In short, the law of excluded middle has to be taken as a clause of the existence principle but not the law of non-contradiction.

The rejection of the law of non-contradiction will of course seem outrageous. To show that a possible solution to the appearance problem is committed to it has been assumed to be a powerful argument against that solution. We must remember however that there are those who have claimed that they can do without non-contradiction as a logical law.[5] So if its rejection were really all that the perspective metaphor committed us to, they at least would not even accept the fact that it implied such a rejection as a good argument against it.

What we have to realise however is that the use of the perspective metaphor does not merely involve us in the acceptance of contradictions; it suggests at the same time that they can somehow be overcome. The incompatible properties of things are to be taken as also compatible, like the different views of an object from different angles and distances. And if it is asked how this is possible, and we do not simply accept it as a further contradiction (this after all would lead to a regress), the only way out is to make a distinction in respect of existence between the incompatible and the compatible, assigning some kind of partial existence to the object in so far as it has incompatible properties and a complete, or less partial, existence to it in so far as its properties are compatible. Treating incompatibles as compatible thus makes it necessary for us to accept that there are grades of existence. This is why the perspective metaphor brings us into conflict with the existence principle.

If we concern ourselves only with contradictions between beliefs, neglecting the case of irreconcilable perceptions, this same transition can be expressed by saying that the idea of the overcoming of contradictions commits us to a doctrine of partial truths. We cannot say of beliefs which are in contradiction with one another that one is simply true and the other false, because the contradiction can be overcome, and then both are incorporated in a more complete truth. Assuming that 'truth' is here being used in the sense in which it is determined by what exists in the world,[6] allowing such grading of truth is to allow the grading of existence, and this, as I have said, is what brings it into conflict with the existence principle.

This idea that we can pass from lesser to greater truths through the overcoming of contradictions is one which is familiar from the writings of Hegel and his many followers. According to Hegel, where there is a thesis and an antithesis, that is, two beliefs inconsistent with one another, it is possible to include them both in a higher synthesis. Neither thesis nor antithesis is to be rejected altogether. They must in some way be true if they are to be synthesised. So that through synthesis we move from partial truths to less partial truths.

Hegel himself admittedly said nothing about perspectives. The idea of partial truths to which his theory about the overcoming of contradictions commits him could well be criticised as falling foul of the existence principle without any reference being made to perspectives at all. It is easy to see however how well talk of the reconciliation of different viewpoints fits in with it. Not only does it inevitably lead to it. It also helps to give it credibility.

One who made explicit use of the perspective idea and combined it with the idea of Hegelian synthesis was Karl Mannheim. His thesis is, as is well known, that all belief systems are culturally determined. Each period, region or class has its own Weltanschauung, its characteristic point of view, and consequently, as he puts it, 'sees the "same" object from a new perspective'.[7] The 'same' here is in quotation marks and this suggests some uneasiness about inconsistent beliefs having the same object, but sameness of object has to be there if he is to use the perspective metaphor as he intends it. It is only if there is one world which can be viewed from different positions that we would normally talk about perspectives at all. The problem is how to combine this with the admission that different socially determined belief systems can be in conflict. And this is where Mannheim, like Hegel, appeals to the possibility of overcoming the inconsistencies through synthesis.

The use of the perspective metaphor in this context is very persuasive. We must now furthermore admit that it is encouraged by certain ordinary ways we have of talking about truth. Even when no mention is made of points of view, we commonly refer to statements people make as having some truth in them, or to opposed theories each having within it an element of truth. No one ever tells the whole truth, the exact truth and nothing but the truth, despite what they may swear to in courts of law. Does not this then in itself tell against our requirement that there are no degrees of truth or of existence? Is not talk of points of view, in fact, something which exhibits to us that there is something wrong with the existence principle?

What we need to show here is that these ways of expressing ourselves are not in conflict with the principle, even though talk of points of view might mislead us into thinking so. The context in which we talk in this way of different amounts of truth is that in which we are concerned either with some set of beliefs or with a certain kind of false belief. We can continue to insist that any given belief which a person holds is either true or false. The point is that when we take a number of beliefs together we can say of that set that it is in some way defective with respect to truth, and when we take a belief which in itself is simply false we can say that it comes near to being true. These cases can be neatly classified under the headings of not being the whole truth, not being the exact truth, and not consisting of nothing but the truth.

Thus the criterion of the whole truth comes into operation when the set of true beliefs held about some object (which may be anything from a particular state of affairs to the whole world) is less than complete. Any such object has a large number of characteristics, and to state the whole truth would be to mention all of them. Though this cannot be done, we may be said to come nearer and nearer to the whole truth the more true beliefs we have about the object. And we may be said to come nearer and nearer to it the more specific the characteristics we include, as when we say not just that there is an electoral swing but that it is a swing of say 7%, or not just that something is red, but that it is cherry red.

The criterion of the exact truth is the one which operates not with sets of beliefs but rather with the closeness of false beliefs to being true ones. If the electoral swing is in fact 7%, it is false to say it is 6%, but this is much nearer to the truth than saying it is 2%. It will be nearer to the truth to say of a cherry red thing that it is vermilion, than to say that it is green.

The criterion of being nothing but the truth comes in, like that of the whole truth, where our concern is with a set of beliefs. Where we have a set of beliefs generally accepted as true, we have to allow that some may be false. This applies in particular where the sets of beliefs are scientific theories from which various deductions can be made. It is well known that in scientific theories there are always some anomalies, that is, statements accepted as true which conflict with deductions from the theory. Since some deductions have turned out false, it follows that the theory as a whole is false. Yet there may be a great deal that is successfully explained by it, that is, many true statements may be deduced from it. So it is natural to say that a theory is nearer to the truth, the greater the number of its true consequences and the fewer its false ones. Neglecting refinements this is Popper's criterion for verisimilitude.[8] According to it one approaches the truth as one eliminates anomalies. Where theories conflict those who hold them may proceed towards reconciliation by a natural dialectic of piecemeal adjustment, or even on occasions major replacement of their constituent statements. These constituent statements of course remain either true or false.

There are various ways then in which certain beliefs or sets of beliefs can be said to have more truth in them than others. This in no way affects the point that where we have inconsistent beliefs we cannot allow that they are both partially true and can be reconciled. Each is either true or false, which is to say that the alleged state of affairs which each says exists either does exist or does not. The use of the perspective metaphor has to be counted as misleading in so far as it suggests otherwise.

BETTER PERSPECTIVES AND WIDER PERSPECTIVES

It will be noticed that I have been moving easily from speaking of partial truth and partial existence to speaking of grades of them. I have been envisaging, that is to say, a doctrine which allows not only two kinds of existence, the partial and the complete, but also degrees of it, the more or less partial. This means that I have taken it to be contrary to clause 2 as well as clause 3 of the existence principle. It is natural to do this, since one assumes that what is partial can be more or less partial, and the Hegelian idea of overcoming contradictions takes it for granted that this is something which can happen in stages.

The question might arise here whether the use of the perspective metaphor in this context allows for such degrees. On the face of it, when an object is looked at from a variety of different positions and angles, it is the same object which is seen. Is not then one viewpoint as good as another? Though this has not stopped the use of the analogy, it might seem that a certain disanalogy is being ignored.

There is no doubt however that we do speak of one viewpoint being better than another, and it is instructive to note the criteria we use. The first of these depends on the fact that anything we look at has a large number of characteristics. From some angles and distances we see more of these than from others and what we see is more specific. There is an optimum distance and angle, for example, from which we can take in the maximum number of details of the ship, both in respect of the number of features observed and their specificity.

Secondly, even though perspectives in themselves do not involve us in illusion, there is no doubt that certain viewpoints protect us against illusions more effectively than others. If we see a ship from far off, as I have pointed out, we are much more likely to take it to be smaller than it is than if we see it from nearby. If we see the proverbial penny at an angle to its face, we are much more likely to take it as elliptical. There is thus an optimum distance and angle for the minimising of errors. And this applies both in respect of the magnitude of any given error and to the number of errors.

Finally it should be noted that both these criteria can be made use of through our looking at the object from various different viewpoints. Moving around it and observing it from different angles and distances is a way of finding out more about it and protecting ourselves from making mistakes.

It should be no surprise that these criteria run parallel to those which we in fact use for gauging the amount of truth in beliefs. A major difference emerges however when we consider what happens when we take up different viewpoints. The extension of the perspective idea naturally leads us to suppose that the more views we have about anything the better. But if these views are incompatible or the beliefs inconsistent, this is in fact not the case. It may be a good thing to consider different views, but in the end we have to reject one or another of any incompatible pair. The perspective metaphor requires that we consolidate them, and that of course is something that could only be done by the overcoming of the contradictions. It thus pushes us from a quite legitimate way of grading beliefs with respect to truth to one which depends on the obscure practice of Hegelian synthesis.

In doing this, we must now point out, it puts the metaphor itself under stress in a rather remarkable way. When you move from one viewpoint to another to find out more about an object, you do not consolidate the viewpoints, you only consolidate the results obtained from each of them. A viewpoint is after all a point, and each point in space is different from every other. What then could it mean to consolidate two viewpoints into a more comprehensive viewpoint?

Yet this is exactly what is proposed when we start to talk of synthesis. Whatever further sense we can give to it, a synthesis has to be regarded as more comprehensive than the contradictory theses which it incorporates. It includes them, so must be a wider as well as a truer truth. In an important respect, therefore, the metaphor does not seem to hold after all. Yet it is a strange fact that, the metaphor being so convenient and persuasive, not even this will shift it. What shifts instead is the use of the word 'perspective' so that it is not even thought odd to talk of viewing things from a wider rather than a narrower perspective. Thus Mannheim is quite happy to speak of the synthesis of various perspectives, and to say that each synthesis attempts to arrive at a wider perspective than the previous one.[9] If we are to keep to the original spatial meaning of 'perspective' such talk of widening perspectives is nonsense. That it is thought necessary should throw further doubt, if it is needed, on the extension of the use of the metaphor.

PERSPECTIVES AND RELATIVISM

Suppose then we drop the Hegelian idea that opposed beliefs can be synthesised into some consolidated point of view. We could then restore the authenticity of the metaphor in this respect; perspectives need no longer be included in wider ones. We would then however be left with the contradictions, and if talk of perspectives is to be of any use at all, something has to be done about eliminating them. We still have somehow to make the incompatible compatible.

There is one final proposal for doing this with which we are already familiar. This is to abandon the idea that there is one world which can be seen from different points of view, and substitute for it a variety of worlds, a different one being seen from each different point of view. This is the relativism about existence which we discussed in Chapter 3.

It has the advantage of eliminating the incompatibilities that appear to exist between the beliefs of different people. The beliefs, being about different worlds, can no longer contradict each other, they are lacking in fact in any possible relation at all. This no doubt is why the metaphor of points of view is so commonly used in the expression of this idea. We have at last found a theory which, in this respect, conforms admirably to the demands of the metaphor. It allows us, moreover, to drop the accretion about narrow and wide perspectives. One perspective need no longer be thought of as wider than another.

We must realise, of course, that this way of making beliefs compatible leaves us in conflict with the existence principle. As an alternative to the Hegelian way of overcoming contradictions, all it does is to move us on from commitment to degrees of existence to commitment to relative existence. In the one case as in the other therefore we are left with an unacceptable consequence. It emerges in fact that there is no way of making incompatibles compatible without rejecting the existence principle. The danger is that in the present case as in the former one, using the idea of perspectives may help to give such a rejection of the principle some credibility.

It is true that in the present case as in the former, the metaphor fails to fit and can only be made to do so by surrendering what is virtually its central point. To make it fit in support of relativism we must leave out the idea that from each of the different standpoints we are viewing the same object. When we speak of different people each seeing a table from a different angle we assume that each is seeing the same table. Yet here we have an account of beliefs which is such, as it were, that no two people ever see the same table. We have reverted from public space to a set of private spaces. And it is a question how much sense there is in speaking of a standpoint in a private space. For after all, in such a space there could not be more than one standpoint.

Such considerations have no doubt induced a certain caution in this last use of the perspective metaphor. The relativism, for example, which is admitted by Stephen Lukes, who makes much of perspectives, is said by him to be only a 'muted relativism'.[10] On the other hand, we should recall that curious use of quotation marks by Mannheim when he referred to the 'same' object seen by each epoch in a different perspective. Since he was very explicit about the perspective metaphor, it was important for him that each should see the same object. Yet at the same time, the quotation marks suggest that he was keeping open a line of retreat to that relativism espoused by so many exponents of the sociology of knowledge who followed him. For how otherwise, Hegelian synthesis apart, could he put his mind at rest about the compatibility of apparently incompatible beliefs? One suspects that his talking of perspectives is a device for having it both ways.

I started this chapter by presenting the perspective idea as a possible way of solving the problem about appearances without introducing third items such as representations. It has become clear however that it cannot cope with illusion and falsity and the incompatibilities between appearances without relying in its turn on one further breach or another of the existence principle. And even then the original idea of a perspective has to be stretched if it is to succeed in making such breaches plausible. Perspectives have their place in the case of the visual perception of objects in space. But beyond that an appeal to them can be dangerously misleading.

NOTES

1 See Thomas Nagel, *The View from Nowhere* (Oxford, 1986). In this book in which intensive use is made of the 'standpoint' metaphor, perspectives are assumed to be determined by circumstances generally, and the spatial origins of such expressions are passed over, despite the spatial reference in the title.

2 In cases of illusory appearance, the word 'see' is sometimes used in such a way that we are said to see the appearance itself, not that which appears. I have not adopted this usage, but have spoken rather of what things are seen *as*.

3 See Ch. 4, p. 56.

4 Kant, *Critique of Pure Reason*, trans. N. Kemp-Smith (Macmillan, London, 1967). For use of this phrase see 'Transcendental Aesthetic', Section I, marginal page A26(B42). G Schräder, 'The Thing in Itself in Kantian Philosophy', in *Review of Metaphysics* (1949) has argued that this is the correct interpretation. He doesn't mention the obvious difficulty.

5 See e.g., Graham Priest, *In Contradiction* (Nijhoff, 1987). He calls his logical theory 'dialetheism'. See esp. Ch. 4, Sects 7-9, where he argues that truth and falsity are mutually exhaustive but not mutually exclusive.

6 On the reasonableness of this assumption, see Ch. 3, p. 39. For further on truth and existence, see Ch. 6, p. 92.

7 Karl Mannheim, *Ideology and Utopia* (Routledge, 1936), p. 243.

8 See K.R. Popper, *Objective Knowledge* (Oxford, 1972) pp. 47-48 and following.

9 Mannheim, *op. cit.*, Ch. III, Sect. 3, esp. p. 135.

10 S. Lukes, 'Relativism in its Place', in Hollis and Lukes, *Rationality and Relativism* (Oxford, 1982) Sect. V.

CHAPTER SIX

THE LAW OF EXCLUDED MIDDLE

I listed as the first implication of the existence principle that we allow no alternative to existing and not existing. And I have taken the assertion of this dichotomy to mean that we must accept, in one of its forms, the traditional 'law of excluded middle'. In Chapter 3 I considered a possible rejection of this law which was associated with a belief in a knowledge-dependent existence as a kind of existence which certain things possessed. It is now time to consider its possible rejection in more general terms—terms which need have nothing to do with knowledge-dependence.

In Chapter 3 I pointed out that, so long as we use the word 'true' in its 'correspondence' sense, the law of excluded middle entails the principle of bivalence, understood as the principle that there are only two truth-values, true and false. It will be convenient if we continue to do this, while keeping in mind that the word 'true' may be used in other ways which lead to questions of truth and falsity being detached from questions about existence. If they are used in these ways they cease to be of interest for us. It is only with the law of excluded middle considered as a principle about existence, and what follows from it so considered, that we will be concerned.

I am assuming here, as I have done previously, that the law of excluded middle can be considered as a principle about existence. Traditionally of course it has been regarded as a basic law of logic in which a certain relation is asserted to hold between propositions—the relation of disjunction between a proposition and its negation. So perhaps some caution is required in associating this with the disjunction I have in mind between existing and not existing.

The association would clearly be complete if every proposition had some truth-value—true, false or whatever, 'truth' being used in its correspondence sense. For then an exhaustive disjunction between the truth and falsity of every proposition, and hence between every proposition and its negation, would be determined by the exhaustive disjunction between the existence and non-existence of some state of affairs in the world which made the proposition true or false. Whereas any proposition without truth-value and any negation of it would have, as I have pointed out,[1] no relevance for the existence of anything whatsoever.

Whether there are any propositions without truth-value is a further question. Propositions about the unknowable constitute one disputed case which we discussed in Chapter 3.[2] We will have occasion to refer to others in due course. For the present all we need say is that in so far as the law of excluded middle applies to them, it is beyond our concern.

But may not this very same question be raised about the law of excluded middle itself? As a law of logic has it a truth-value? Can the assertion of the dichotomy between existing and not existing itself be determined as true or false by how the world is?

A natural answer to this question is of course that if we accept this law it is because it is true, in the correspondence sense. And if we take it to be true, we assume we must regard as false all those proposed alternative systems of logic which have abandoned it. We must remember however that there is a view about logic which takes the acceptance or rejection of such systems as a matter not of truth but of comparative convenience.[3] On this view the law of excluded middle will lack truth-value, and will have nothing to say about existence. What we require therefore if the law is to be part of our existence principle is that we do not take this view of it. We must take it to be making a claim about the general structure of the world, and we must take any alternative logic which challenged it as involving a denial of this claim. If the choice between such logical systems were a metaphysically arbitrary one, being between different but equivalent frameworks for describing the world, we would be entitled to lose interest in whether the law of excluded middle was accepted or not.

It is doubtful however how seriously we can take this view of logical principles. Once a criterion for choice is mentioned, such as convenience, it is hard to distinguish it from the criterion of truth. If it is asked why it is more convenient to use one logical framework rather than another, the answer is likely to be, say, that it is simpler or more coherent, or even better fitted for dealing with certain facts.[4] And these after all are common tests for truth. Hence in cases of rivalry between a logical system which includes the law of excluded middle and one which does not, a consideration of the truth of the law gets brought within our scope. The consequence of this is that it becomes a principle about existence applicable to any proposition which itself has truth-value.

In requiring in this way that it be taken as true, we should keep in mind that it has a special role to play which distinguishes it from other laws in a logical system. We are concerned with it not only as a very general statement about the structure of an existent world, but also as a requirement imposed by our basic concept of existence. Those who put forward an alternative logical theory which eliminates it—say, an intuitionist theory—are not only differing about logic but are challenging the existence principle. This makes no difference to the claim that it is true but it does make a difference to the way in which it needs to be defended. It has to be defended not as a logical law but as an essential clause of the existence principle.[5]

With this understanding of our concern with the law of excluded middle, we can now consider some of the ways in which it has been challenged.

PRESUPPOSITIONS AND LIMITS

There is one whole class of these which we can now deal with fairly shortly. These are those in which it is claimed that the existence–non-existence dichotomy or the truth-falsity dichotomy is context-dependent and only applies within certain limits. We have already discussed an argument of this kind in Chapter 3, an argument in which the law of excluded middle has been challenged in connection with the anti-realist theory of the knowledge-dependence of existence. On this theory statements beyond the limits of possible knowledge while being genuine statements are neither true nor false, and hence, assuming the correspondence sense of 'truth', might be thought to be concerned with what neither exists nor does not exist.

That argument, we saw, failed because even if statements about the unknowable were neither true nor false, which many would deny, all that could be said of them would be that they lacked truth-value, and from that nothing about alternatives to existence could be inferred. What we need now to realise is that this consideration can be applied more generally to any case of alleged context—dependence of truth and falsity. It has sometimes been said, for example, that the true-false dichotomy only applies within some universe of discourse, and is inappropriate when an attempt is made to extend it. Given that one is talking of things coloured, one may insist of one of them that either it is red or not red. But if you extend this to things to which colour-predicates do not properly apply, such as micro-particles, or minds, and say of one of these that it is red, this cannot be said on this view to be false any more than it can be said to be true. All that can be said is that you have made a category mistake.

If we accept this, it still remains irrelevant for questions about existence. As before, there are two possibilities. On the one hand it may be said that the word 'true' is simply being used in a different sense when restricted to the universe of discourse. When it is used in the correspondence sense, we are committed to judging of truth and falsity outside the specified limits. There is in fact no state of affairs in the world describable as that of a mind's being red, and the statement that one is therefore on this view simply false. This is so obvious that we may not bother with it, and this may lead us to restrict the use of 'false' to cases where there is some genuine issue.

On the other hand, if we deny that we are here using 'true' and 'false' in a restricted sense, and insist that, in their correspondence sense, the statement that a mind is red has no truth value at all, it follows that there is nothing which could make it true or false or with any other truth-value. Any putative alternative to red minds existing or not existing could have no truth-value-making function, and so any argument from its having such a function would lapse.

What is said here about the limits imposed by universes of discourse may be extended to include the limits imposed by presuppositions of all kinds. Where something is presupposed in the making of a statement, it has often been insisted, that statement is neither true nor false unless the presupposition is true. To say of a lifelong non-smoker that he or she has stopped smoking is to say something the truth or falsity of which does not arise. And, as we know, an important case of this is that in which we presuppose the existence of something to which we attribute some property, as when we say, for example, that the present king of France is bald, or that Zeus is all-powerful. Where the subject-term lacks a reference the statement, we may say, lacks a truth value, and bivalence fails.

It is a familiar fact that Bertrand Russell denied this, claiming that the presupposition has to be included as a conjunct in the statement. It follows that where the presupposition is false (there exists no king of France), the statement as a whole has a perfectly definite truth-value, it is false also. 'Truth' is here being used in its correspondence sense both for the existential presupposition and the statement as a whole, and yet no challenge to bivalence arises.

What are we to say, however, to those who have insisted that there is a challenge? Once again we may point to our two alternatives. In this case, it seems very natural to adopt the first and say that where truth-value is made to depend on presuppositions, 'truth' is being used in something less than the correspondence sense. As has been pointed out by Peter Strawson, a well known critic of Russell's account of this matter, which account you adopt really depends on the direction of your interest. If you are interested in actual speech situations, you will find people not including the presupposition within the scope of their statement, and hence the truth-value gap theory is natural.[6] And what is this than to be speaking of truth in a restricted way, which detaches it from questions of existence?

As far as our concern is with existence, however, it does not matter whether we accept this account or not. So long as they lead to the acceptance of truth-value gaps and not to alternative truth-values, statements made under presuppositions can have no implications for existence whatever.

A final case of this kind to mention is that of statements of fiction. In story-telling, as distinct from biography, we are of course not interested in whether our characters exist or whether what we say about them is true. There would seem therefore to be no harm in adopting the Russell alternative and classifying them as false, without any derogatory intent. But if we choose not to do this and say the statements in the story lack truth-value, we may proceed as before and claim that questions of truth-value are being detached from questions of existence.

Admittedly there is a difference of which we should take note when we come to works of fiction. This is that they consist of many statements which hang together in a connected narrative, giving a complex picture of a whole human situation. This has been a reason, not for classifying them as neither true nor false, but for saying that they have their own truth within their universe of discourse, which does not apply to what is beyond. Thus we may want to know whether Hamlet was really a procrastinator, or whether his father's ghost was not just a figment of his imagination, and such questions simply do not arise outside the limits of the play.

There is of course no harm in discussing the internal situations described in a work of fiction in this way. But it should be clear that when we talk of the existence of such situations we are either being metaphorical or doing what I suggested in Chapter 1 is often done, using the word 'exist' not in its basic sense, but merely to indicate that something is an item in some universe of discourse.[7] Otherwise, it should be pointed out, we would be landing ourselves into the position of envisaging innumerable kinds of relative existence, each belonging to one of the innumerable works of fiction. This would involve a challenge to the existence principle of a much more extreme kind than a challenge to the law of excluded middle.[8]

SHADOWS AND APPEARANCES

Let us now turn to cases where it is claimed that there are truth values providing a positive alternative to truth and falsity. Such claims cannot be dismissed as irrelevant in the same way as those for truth-value gaps, since they present us with a corresponding positive alternative to existing and not existing. It is in these cases that the traditional reference to a 'middle' tends to have point since it is common, though not essential, to envisage such alternatives as somewhere in between existing and not existing.

I have already referred in Chapter 1 to Plato's conception of the sensory world as lying somewhere between the realms of non-existence and full-existence.[9] In this case, as I said, it might be doubtful whether he intended a breach of the law of excluded middle, since he might have intended to grant a kind of existence to sensory things, but one which was different from, and inferior to, that granted to the forms. And it must be admitted that it is always dangerously easy to slide over from one of these ways of challenging the existence principle to the other. If we reject one of these ways, however, we need to close a loop-hole by rejecting the other. I will assume therefore in what follows that in all doubtful cases what is being spoken of is an alternative to existing and not a kind of it.

The trouble with alternatives to existing is that it is often not easy to make clear to oneself what the alternative existence value is supposed to be. Sometimes when it seems clear enough, we find that this is because the words 'exist', and more particularly 'real' are not being used in their basic sense. As I mentioned in Chapter 2, Plato was inclined to identify the existent with the independent the unchanging or the good and for that reason to present the dependent fleeting and imperfect things of sense as not existing.[10] When this kind of thing is done, as I pointed out, any challenge to the law of excluded middle can be dismissed as due to misunderstanding.

Sometimes however the appearance of one's knowing what is being talked about is given by the use of some supposedly helpful metaphor. When Plato spoke of the status of the sensory world he often used the metaphor of shadows, a metaphor which is made full use of in his story about the cave.[11] And the idea that a shadow is not an existing thing, and yet not nothing, can be very persuasive. We should therefore look at the concept of a shadow a bit more carefully.

In the concept of a shadow there are of course two ingredients, the absence of light due to an obstruction in front of the light source, and a dark part of some surface surrounded by a lighter one. The first of these refers to something not existing, namely, light beams in a given region and their reflections from a given surface. The second, on the other hand, refers to something which exists alongside other things in the world around it, namely part of a surface darker than that surrounding it. If you put together these two items, the non-existence of something and the existence of something else, you do not get something which is between existence and non-existence. Yet I suspect that it is the fact that there are these two ingredients that gives rise to the illusion that we have here some sort of compromise. To this we should perhaps add that a shadow, as a dark patch, is a dependent entity, caused by the obstruction to the light source, and therefore might also fall to Plato's assimilation of existence to independence. Whatever the explanation, however, it should be clear that the idea of a shadow as a 'middle' to be taken account of as a third alternative has no basis whatsoever.

Another way in which to attempt to make clear to ourselves what an alternative existence value might be is to refer to the status of appearances. This case may be said to be of more significance than that of shadows, since we are not here trading on a metaphor. In Chapter 4 we discussed the problem raised by a feeling on the one hand that how things appear has to be distinguished from how things are and a feeling on the other hand that the appearances could not be simply eliminated but had somehow to be taken account of. This problem might be thought to be easily solved not only, as I there suggested, by attributing to appearances a different kind of existence, but instead by treating appearing has having some sort of middle status between existing and not existing.

This claim for a separate existence-value for appearances clearly has to be met in the same way as the same claim for such things as shadows. What we need to do is to give an account of the appearance situation in terms which require only the two values. This can be done in ways which run parallel to those adopted in Chapter 4, where we were dealing with the interpretation of appearances as requiring a separate kind of existence. We can point to two well-known ways of avoiding the need for any middle value. One is to treat appearing as no more than a relation between two existents, the other is to regard appearances as entities existing in their own right. In neither case is any reference to a third value required. In discussing the merits of these alternatives, we saw some good reason for adopting the second especially when the issue is extended to intentional objects in general. But this takes us beyond the present issue. What is important is that, while the idea of appearing being in the middle between existing and not existing is an attractive one, any serious discussion of the issue has to be in terms of theories which avoid it.

INDETERMINATE OBJECTS

We come now to another important way in which the gap between existing and not existing might be thought to be filled, and that is by things and states of affairs which are said to be indeterminate. In Chapter 1 I mentioned as a possible additional requirement for the concept of existence that all existence is determinate. What can now be pointed out is that the reason for not including it is that, given standard logical principles, it follows from the first requirement which precludes alternatives to existing and not existing. If we consider the concept of indeterminacy it will become clear that alleged indeterminate objects cannot be said either to exist or not to exist. If there is indeterminacy in the world it requires a third existence-value, statements about indeterminate objects being neither true nor false but having some correspondingly indeterminate truth-value.

Of alleged indeterminate objects, there have been various kinds proposed. First, there are phenomenal objects such as my visual impression of a speckled hen which, though it has speckles, may be said to have no precise number of speckles. Secondly, there are vague objects such as a collection of grains of sand which is neither a heap nor not a heap but a borderline case between the two. Thirdly, there are microphysical objects such as the electrons or protons which, according to some interpreters of quantum theory, have some position within a region but no particular position. We should also mention future objects such as tomorrow's sea-battle made famous by Aristotle, but the indeterminacy said to be involved in these is something we can leave to the next chapter, to be considered in connection with other questions about time and existence. I will conclude this chapter by saying something about the other three.

But first we need to try to get clear what this indeterminacy consists in. A first thought might be that an indeterminate object is one which has some determinable property but simply lacks any of some range of determinate properties which fall under it. On such an account of it, the hen impression would be speckled but it would not have any specific number of speckles at all, there would be a collection of grains of sand but it would neither be a heap nor fail to be a heap, the electron would be in a region but at no place in the region. If this were all there was to it, indeterminacy would be a very queer situation to be in, but it would involve no violation of the existence principle. It would consist in a certain state of affairs existing, that of something having a determinable property and certain other states of affairs not existing. The oddity would be that the others are certain determinate ones. This would involve, no doubt, a challenge to the law of contradiction, but not to that of excluded middle.

As an account of indeterminacy however this clearly will not do. If someone says that the hen-impression has exactly 124 speckles, we do not want to say that this is simply false, and that it is simply false that it has 123, 122 or any other number. What we want to say is that, though it has some number of speckles or other, it is indeterminate whether it has some particular number. That it has 124 speckles therefore is neither true nor false; it is merely indeterminate whether it has them or not. Similarly it is indeterminate whether some collection of sand-grains constitutes a heap or whether the electron is in some particular position in the region. In the case of the electron, this is made particularly clear when those quantum theorists who deal in these matters are prepared to speak of the probability—the objective probability—of the particle's being in a particular part of the region.

A belief in indeterminate objects, therefore, offers a challenge not to the law of contradiction but to the law of excluded middle. A third alternative is being proposed to something either being there or not. And this alternative can only be characterised by its being said to be indeterminate. Objective indeterminacy thus remains as a primitive concept. There seems no way in which we can produce a reductive account of it. But like all alternatives to existing and not existing, it gains colour from analogies on which it draws. We are familiar with the indeterminacy of our knowledge as when we are uncertain which position the electron is actually in. And we are familiar with the vagueness of the expressions we use. Ideas from such sources are no doubt drawn upon when we make the indeterminate into an addition to the existing world. Similarly when vague objects are described as fuzzy, as is sometimes done, a spatial metaphor is being introduced which suggests quite wrongly that indeterminacy is like the gradual fading off of indistinct boundaries. It would be a mistake to let it sail in under such false colours.

However difficult it may be to say what objective indeterminacy consists in, one thing is clear and that is that unless we abandon some other part of our logic, positing it brings us into conflict with the existence principle. And we may well ask what would drive anyone to take this step. The reasons are very different in the different cases we are considering. But in each case there is, as we would expect, a problem to which a resort to indeterminacy appears to provide an easy solution.

In the case of sense-impressions, the reason for resorting to indeterminacy has been a very simple one; it is supposed to be a fact of immediate observation. About such impressions no mistakes are possible, and it is a fact that one observes the speckled hen without observing the number of speckles. It has seemed that the only way to account for this is to say that it in fact has an indeterminate number of speckles.

In the case of microphysical particles there is a similar concern with the explanation of established facts, but in this case they are not facts of immediate observation but certain puzzling facts which have come to light in the course of physical enquiry. These have appeared to be inexplicable on any other hypothesis than that certain particles have no precise combination of position and momentum at any moment.

In the case of vagueness, the problem is of a very different kind. It is one of how to account for the quite general vagueness of the linguistic expressions we use in describing the world. When we say that something is a heap, or is red, or that someone is bald, it may be argued that there must be some corresponding state of the world which makes such statements true or false. But these expressions are vague and there are borderline cases where it seems unreasonable to insist that they are true or false. We have here a third truth-value, which we may call 'indeterminate', and this may be thought to require a third existence-value, that of indeterminacy in the world which confers this truth-value upon them.

Though these reasons are various, they all point to the same conclusion, that there is indeterminacy in the world. The conclusion is, to say the least, counter-intuitive. When people affirm a disjunction of alternatives, such as that something is either a heap or not a heap, or that a particle is in some position in a region or another, and then say it does not follow from this that any of the alternatives must be accepted, we will be inclined to think that there is something wrong with their logic.

At this point however it may be said that this is just the trouble, that we are accepting without question certain principles of classical logic, and that what we need to do is to modify our logic in a way which fits with the conclusion.

One way of modifying one's logic to fit with the conclusion would of course be to abandon the law of excluded middle.[12] But this, as I have pointed out, would mean not only changing one's logic but challenging the existence principle. So it merely brings us back to the previous point of the conflict between indeterminacy and that principle.

There has however been another suggestion for modifying one's logic which if accepted would avoid that conflict. This is that we abandon not the law of excluded middle but the distributive law according to which any object which has a disjunction of properties either has one of the properties or has the other.[13] If we did this, we could retain the dichotomy of either existing or not existing, and insist instead that all that is required for indeterminacy is that the disjunction between alternatives be treated as ultimate.

I think we must agree that the idea of disjunctions being ultimate does not affect the concept of existence any more than the admission of contradictions. But like that admission, it would be a very high price to pay. The distributive law is used throughout all argument, and it may be doubted whether those who claimed to abandon it would ever be able to do so consistently. In so far as the standard truth tables give the meaning of such logical connectives as the disjunction sign 'V', it may be said that the required result is only achieved by changing the meaning of the connective.[14] Such a way therefore of reconciling indeterminacy with the existence principle should be treated as a very last resort. We should rather return to the question whether indeterminacy itself can be avoided, despite the reasons we have noted that have been given in its favour.

WAYS OF AVOIDING INDETERMINACY

(a) Phenomenal Objects

The case of the alleged indeterminacy of phenomenal objects is the easiest one to deal with for we have here two alternatives. Either we abandon the theory about appearances which requires such items, or, accepting the theory, we argue that the indeterminacy conclusion does not follow. As we have noted, there are those who reject the theory, and in doing this the claim for indeterminacy which goes with it. In fact the argument has on occasions been reversed, and the presumed indeterminacy of immediately given data has been used as an argument against the view that they exist as separate items.[15] Whether the rejection of this representationalist view is tenable is something we have already discussed.[16] But that it would be a way of avoiding indeterminacy could be counted as a point in its favour.

One might however question the need to abandon the idea of there being immediately given data in order to avoid the conclusion that there are indeterminate entities. We should of course avoid the temptation to do what some phenomenalists have suggested, and say that while all the data are determinate we could make mistakes about them.[17] For while that would solve the problem, it would only be at the expense of defeating the purpose for which such items were originally posited. We would invite a regress if appearances were allowed to appear other than they are.

There is however another way of solving this problem which does not run us into such difficulties. Let us agree, it may be said, that there are immediately given data and that they do not appear other than they are. But this does not stop these data having other features which we do not experience. We can allow, in fact, mistakes of omission, if these can be counted as mistakes at all. We can always insist therefore that the speckled hen appearance has some determinate number of speckles, it just happening to be the case that we have not taken notice of them.[18]

We may be in dispute about which of these two kinds of theories we should adopt. The important point for the present question is that whichever we adopt a way is left open for avoiding the view that there are indeterminate entities.

(b) Vague objects

About vagueness there has been much controversy. Fortunately in the course of this various proposals have been made which would enable us to avoid the admitted vagueness of our language having any existential implications. What we have to insist on is that however we choose to describe it any alleged state of affairs must either exist or not. This means that in borderline cases anything must either have the specified property or not. The cut-off point may well be a matter of convention, which may differ for different people. There is no doubt a range of possibilities about what constitutes a heap when grains of sand are added. But it remains the case that if we are going to talk of the world in terms of heaps or their absence, there must be some cut-off point.

Given this requirement, one thing we might say is that all vague statements are false, and hence no question arises about alternative truth-values in borderline cases. At the borderline, on such a view, it is false to say the collection of sand-grains is a heap, because there are in the world no such things as heaps. The only true statements are precise statements and the only objects are those which these statements are about.[19]

This no doubt would solve our problem, but in view of the fact that most of the claims we make about anything are vague, it would have a devastating effect on our views about the relation of our language to the world. Surely, we say, if the sand is piled up high there is no doubt there exists a heap; if there is only a handful of it there is not. It is in the neighbourhood of the borderline that the question about the truth or falsity of vague statements arises.

This brings us to another proposal which would take account of this consideration. According to this vagueness is not so much a matter of falsity as of ignorance. Given that there is a definite cut-off point, evidence as to where it is becomes more and more tenuous as we move towards the centre. There are clear cases where we know that our statements about heaps are true or false, but there comes a point at which we will refuse to commit ourselves. So we may say that it is this area of uncertainty about the truth of a statement which makes it vague.

There are those however who are unwilling to accept this reduction of vagueness to ignorance of the precise. So what they have to do is reconcile the truth of vague statements with the absence of any vagueness in the world which these statements describe, even at borderline. Fortunately we can point out a way in which this can be done. We may take our cue from Bertrand Russell who many years ago put forward the simple idea that because of vagueness the relation between our linguistic expressions and the world was a one-many rather than a one-one relation.[20] To require vague objects to make vague statements true, we might say, is to give an unduly narrow interpretation to the use of 'truth' in its correspondence sense, one in fact which makes it into one of Russell's one-one relations. Disjunctive statements, as we know, are not made true by objectively indeterminate disjunctive facts, unless, that is, we reject the distributive law. They are made true by the actual existence of one or other of the disjuncts. Similarly, we may say, vague statements are not made true by vague facts but by some collection of determinate facts. If each of the determinate facts were such as to make the vague statement true, we would count it as true, though not in the untenably narrow correspondence sense. And this applies at the borderline as much as anywhere else. The difference from the case of disjunctive statements is of course that we require not one determinate fact from a collection of alternatives but a conjunction of them all. But it is still the case that we require nothing but determinate facts.[21]

It is clear then that there are various ways in which an account of vagueness can be given without requiring the rejection of the law of excluded middle. In particular the last of them, though only presented here in outline, is one which, if followed up, could well prove an effective way to solve the problem.

(c) Microphysical objects

The case of the alleged indeterminacy of microphysical particles is not such an easy one to discuss, because the offering of alternatives to indeterminacy depends on the understanding of complex matters of modern physical theory. If a physicist claims that the indeterminate position of particles is the only way to account for certain recently discovered physical facts, one would, it would seem, have to be deeply involved in the subject to challenge the claim.

It is useful however to make a distinction between the effective down-to-earth operation with the tools of quantum mechanics, such, for example, as the mathematics of vector spaces, and the more speculative considerations about their interpretation.[22] It is with the latter that we are concerned, and it is well known that there are interpretations which challenge the indeterminacy view.

In support of such challenges, we should keep in mind three things. The first is that it is not at all clear, as we have seen, what the objective indeterminacy, or objective probability, of the position of any particle could consist in. The second is that it is quite clear that an indeterminately positioned particle could never be observed. There is no parallel here to the case of the speckled hen. All quantum theorists agree that whenever anything is measured, it is in a determinate position. That it has been indeterminate is a hypothesis based on determinate data. And the third, which we should not forget, is that the indeterminacy hypothesis is in conflict with the existence principle which we have argued is required as a presupposition of all theorising. All of which suggests that we should not take the indeterminacy hypothesis as one among others on an equal footing, but rather something to be avoided if we possibly can.

One obvious way of avoiding it would of course be to say that the whole problem is due to the limitation of our knowledge. Though the properties of particles are determinate, we could claim, our capacity to discover what these properties are is limited, not only in practice but in principle. On this view there are 'hidden variables' which, if posited, would account for the surface phenomenon. Such a view encounters difficulties. There are experiments such as the 'two-slit' experiment, the results of which have been claimed not to be explicable in this way. There are however those who believe such difficulties can be overcome.[23] And there is always the consideration that the indeterminate particle theory is a theory, and, as the history of science shows, theories can be replaced with the progress of science.

We should remember that quantum theory itself has made a great difference to our view of the physical world. It has brought us up, for example, against the possibility that the ultimate laws of nature are probabilistic in character and that there is action at a distance. Such changes in theories do not affect the existence principle. But they are substantial changes nevertheless. They should encourage us to look for a further change in respect of a theory which does affect that principle, the theory that there are such things as indeterminate particles.

Quite apart from these general considerations, there is a further twist which might be given to the indeterminacy theory which deserves mention. It is a curious fact, on this theory, that once you measure the position of a particle, that position becomes determinate. One has to ask why this should be. It makes it look as if the existing world of micro-objects is in some way dependent on measurement. The indeterminacy of those objects which are outside its scope could then be said to consist in statements about them being neither true nor false.[24]

This might be called the 'anti-realist' solution to the problem. What it does is to treat existence as knowledge-dependent. In doing this it can, as we have seen, avoid a challenge to the law of excluded middle, considered as we have been doing as a principle about existence. For, as I have argued, in asserting knowledge-dependence it requires no more than truth-value gaps for statements about what lies outside the scope of measurement. Statements about unmeasured particles having no truth-value, there would then be nothing in the way of alternatives to existing and not existing which could be assigned to them.

From the point of view of the existence principle, this is of course of no help, for it merely replaces one breach of it by another. Resorting to some non-causal measurement-dependence may by-pass indeterminacy but it does so by positing instead a special kind of existence for the measured objects. If we are to avoid the disastrous consequences of admitting indeterminacy in the world, we should not be led into this kind of cul de sac. We should rely rather on the progress of science for the discovery of other solutions.

NOTES

1 Ch. 3, p. 40.

2 Ch. 3, p. 41.

3 See Ch. 2, p. 22.

4 When Susan Haack, *Deviant Logic* (Cambridge, 1974), Ch. 2, advocates what she calls a 'pragmatist' view of logical laws, we find that she moves from the criterion of 'convenience' to others, such as 'economy, coherence and simplicity'. It turns out in fact that what she has in mind is merely that no statement (even of logic) is, in Quine's words, 'immune to revision'. This, which we may accept, does not rule out conflict between logical systems in respect of truth.

5 N. Rescher, *Many-Valued Logic* (McGraw-Hill, 1969), Ch. 3, pp. 232-234, appears to maintain that we can accept the law at the meta-logical level while not necessarily accepting it as a constitutive part of a logical system. He contrasts the 'Regulative Absolutism of Meta-Logical Principles' with the 'Constitutive Relativism of Logical Theses'. Though the distinction is well-made, it is surely one in the kind of defense required, not in truth.

6 See P.F. Strawson, 'Identifying Reference and Truth Values', in *Theoria* (1964): reprinted in *Logico-Linguistics Papers*, Paper No. 4.

7 See Ch. 1, p. 2.

8 I neglect here another use of the idea of 'truth in fiction', that in which a work of fiction may illustrate and make us aware of various subtle truths about human nature. I do this because such truth is clearly truth in the correspondence sense, and what are being brought to light are facts about the existing human world. No issue about bi-valence, therefore, arises.

9 See Ch. 1, p. 5.

10 Ch. 2, pp. 16-17. Note that I was there interpreting Plato's distinction as one between degrees of existence.

11 *Republic*, pp. 514-519.

12 Recall Quine's claim that logical principles are not immune to revision and his mention in this connection of the proposal to abandon this law to simplify quantum theory. See W. van O. Quine, 'Two Dogmas of Empiricism', as reprinted in *From A Logical Point of View* (Harper Torchbook, ed., New York, 1963) p. 43.

13 For this proposal, presented for the case of quantum logic, see H. Putnam, 'The Logic of Quantum Mechanics', in *Philosophical Papers*, Vol. I (Cambridge, 1975). The formula for this law could be presented as: a. b v c → a. b v a c. Alternatively, to fit the case of the heap, a. bv – b → a. b v a – b.

14 See the paper by M. Dummett, 'Is Logic Empirical?', in *Truth and Other Enigmas* (Ch. 16) (Duckworth, 1978) for criticism of Putnam's proposal along these lines.

15 See Winston Barnes, 'The Myth of Sense-Data', in *Proceedings of the Aristotelian Society*, 1944-5. For an argument of this kind, David Armstrong, *Perception and the Physical World* (Routledge, London, 1961) Ch. 4, Sect. 2 and Ch. 10, Sect. 5 has also taken this line.

16 See Ch. 4, pp. 62-64.

17 See A.J. Ayer, *The Problem of Knowledge* (Penguin, 1956) Ch. 2, Sect. VI, pp. 65-68 for the view that such mistakes can be made, the appearance of one line, for example, being wrongly taken to be longer than the appearance of another.

18 See C.D. Broad, *Scientific Thought* (London, 1923) p. 244 for the making of this point. See also Frank Jackson, *Perception* (Cambridge, 1977) pp. 115-117.

19 See, e.g., for views of this kind P. Unger, 'There are no Ordinary Things', in *Synthese*, Vol. 44 (1979); Samuel E. Wheeler, 'On That Which is Not', in the same volume; Mark Heller, 'Vagueness and the Standard Ontology', in *Nous* (1988).

20 See Bertrand Russell, 'Vagueness', in *The Australasian Journal of Psychology and Philosophy*, Vol. 1 (1923).

21 This, as I understand it, is the theory presented by Michael Dummett, 'Wang's Paradox', in *Synthese*, Vol. 30 (1975), and by Kit Fine, 'Vagueness, Truth and Logic', in the same volume and called by Fine the 'super-truth' theory. As expressed by Fine 'a sentence is true if it is true for all ways of making it completely precise'.

22 See Peter Forrest, *Quantum Metaphysics* (Blackwell, 1988) for a careful drawing of this distinction.

23 E.g. originally Einstein, later Karl Popper, David Bohm.

24 There are elements of this theory to be found in what is called the 'Copenhagen interpretation' of quantum theory, in particular as expounded by Niels Bohr.

CHAPTER SEVEN

TIME AND EXISTENCE

So far in discussing the concept of existence and its implications I have made no mention of time. The question now arises whether the core concept of existence has temporal implications which we have overlooked and, if it has, how this might affect the existence principle. Does the need to ask *when* something exists prevent us in any way from giving a straight answer to the question about its existence, yes or no?

TIME-INDEPENDENT AND TIME-DEPENDENT EXISTENCE

We can begin with a consideration about language. As we all know, when we use verbs like 'to be' or 'to exist' they always have a grammatical tense. Basic to these tenses are the present, past and future; we distinguish between what exists, what existed and what will exist. There are of course also more complex tenses, such as the perfect ('has existed') future perfect ('will have existed') and so on. But since these are compounded out of the basic three, we may simplify by ignoring them.

The obvious point in our using these different tenses is to distinguish between things and states of affairs which are in existence at the time of our speaking and those which are in existence at times before and after this time. If this is all there is to it, it will be clear that our use of tenses has no implications about the existence of things or states of affairs. We assume that anything that happens at any time simply exists. When it exists is a further fact about it. And in so far as we take our time of speaking as a point of origin—the present—we can fix other times for our convenience by reference to this. What is said to be happening now will of course vary with the time of speaking, so that statements about what happened and what will happen will have what is called a 'token-reflexive' character varying in truth-value according to when they are made.

If this were all there was to it, there would be no more to be said. But it is common knowledge that many have thought that this is not all there is to it. If we are to take time seriously, it is said, we cannot treat the present, what is happening now, as some arbitrary point in a series of events, fixed by when in that series any given person is happening to be talking or thinking about the world. The present has rather to be regarded as a real fact about the world itself—a watershed in the nature of things, flanked by past and future.

This real fact might of course be envisaged as merely an additional fact about the world which would be quite consistent with the basic fact that everything in time existed. All that might be being said is that there is always one item in the time series which not only exists but is also the present moment, presentness being a feature which shifts continually from one item to the next. We should keep this possibility in mind. But there are those who would regard this as insufficient to account for the central reality of the present. This central reality, on such a view, requires that it exists in some way in which past and future do not. Existence becomes present existence, and this puts us on the way to the view that there is an intrinsic time reference in the very concept of existence.

This connection of existence with presentness is made plausible by the way we use tenses. When we speak of something existing we use the present tense. Words like 'existed' and 'will exist' are derived from the present tense use. There is of course a time-independent use for the present tense. There has to be since otherwise we could not even make sense of the Platonist's claim that numbers exist or that universals exist. But this is simply because the objects referred to are non-temporal. For anything that exists in time, saying that it exists carries with it the idea that it is present. Even of everlasting atoms we do not say simply that they exist in the time-independent sense. We need to add, if we wish to refer to their everlastingness, that they have existed forever and will exist forever.

It may be said of course that this is a merely verbal matter. There is a tenseless use of the verb as well as a tensed one, and the fact that the present is used for it is typical verbal economy. In some languages such as English it sounds odd to adopt the time-independent use for temporal things, but we could easily devise one. We could simply point out that the word 'exists' has a special sense of 'exists now' as well as the general sense which makes no reference to time. We could then claim that the general sense was the basic sense, and the one we have in mind when we state the existence principle.

The question however remains, what would justify this last claim. There is nothing in word-usage itself to determine which sense is basic. Might it not be that 'exists' in the sense of 'exists now' refers to the core concept of existence, 'existed' and 'will exist' being derived from this, and the timeless use, for temporal things, being equivalent to the disjunction 'existed, exists or will exist'? The possibility remains open therefore that presentness, or something connected with presentness, is central to the concept.

It is important to notice however that merely asserting a connection between existing and being present does not take us as far as that. Those who assert it, that is to say, are not necessarily committed to including this time-reference in their concept of existence. All they may be doing is offering us an alternative metaphysical theory about what exists, to replace the theory that every item in the time series exists. In offering us this theory they may be assumed to be retaining the time-independent concept of existence as basic. With this concept in mind they may be claiming that not every alleged item but only certain of them exist. This need in no way affect their meta-metaphysical assumption about the concept of existence.

Whether anyone has actually held such theories, with the time-independent concept of existence kept intact, is not an easy question to answer. But there are many who have given the appearance of doing this. The simplest of such theories is of course that only present items exist and nothing else, past or future, at all. This certainly fits a common way we have of talking when we use tenses. We do say of things past that they have ceased to exist, and of things future that they have not yet come into existence. If 'exist' were being used in the special sense of 'exist in the present' this would indeed be nothing more than a tautology. But it does not look like a tautology. Augustine, I am sure, was not regarding it as such when he said that 'neither things to come nor things past are'.[1] Nor presumably on first impression was Arthur Prior when he said that the present tense was vacuous, and to say of someone that he is eating his breakfast now is to say the same thing as that he is eating his breakfast, the indication of its presentness simply being redundant, like the phrase 'It is true that' before 'he is eating his breakfast'.[2] As another more recent writer has put it, 'What is not happening now, but was or will be happening at another time, is not happening at another location, it is not happening at all.'[3]

It must be admitted that, taken as a straight metaphysical theory, this theory of the sole reality of the present would be very hard to maintain with any consistency. For one thing there is the problem, already noted by Augustine, about how long the present lasts. If we take it to its natural limit, it appears to be a knife edge which separates past from future. And how can such an edge exist if there is nothing to separate?

This may be part of the reason for the softening of the theory to allow for the existence of more than the merely present, but without taking in the lot. We may note here two such more moderate versions. The first is one which switches from the existence of events to the existence of enduring things and holds that enduring things exist so long as they overlap the present. The second, and for us the more important one, is that which makes the present an edge of the existing past, but denies existence to the future. This is a theory which has a very positive appeal, in that we are inclined to think of what has happened as having been incorporated into a world, this being a world which grows, continually enlarging itself with the addition of new items as time moves on. In Bergson's colourful language, real time on such a view is to be thought of as 'the continuous progress of the past which gnaws into the future and which swells as it advances'.[4]

Despite their avoidance of the knife-edge problem, however, and despite the special plausibility of the second version, any such modifications continue, along with the simple version, to be very difficult to maintain as theories about what exists. We need only consider the effect which the simple denial of the existence of any item in the time series must have on our views about historical and predictive-scientific knowledge. If the past does not exist, there would be nothing for a historian to find out about, and a statement, for example, asserting the existence of a battle at Waterloo in 1815 would be false. If the future does not exist, the same would apply to anyone who affirms the existence of a sea-battle tomorrow, or of an eclipse of the sun in two years time. As in the case of dragons, we could imagine what they would be like, but to think we could make discoveries about them would be inconsistent.

In view of such difficulties, we may begin to wonder whether such theories have ever been seriously maintained. This is the point at which we should consider whether those who have appeared to maintain them have in fact been doing something more. For such theories are based on the assumption that we are working with the simple time-independent concept of existence which conforms to the requirements of the existence principle. And it may be thought that the difficulties can be avoided if we drop this assumption.

This line of thought may be followed up in two very different ways. One of these is a general one which starts from the sole existence of the present. The other is a more specific one which starts from the non-existence of the future. Let us consider these in turn, starting with the specific problem about the future.

INDETERMINACY OF THE FUTURE

The issue about the existence of the future has commonly been presented in terms of truth. If future states of affairs exist, any statement concerning their existence must be either true or false. If they do not exist, all such statements will be false. One way of avoiding this second unpalatable alternative without adopting the first is to retreat into the middle and say that such statements are neither true nor false but have an indeterminate truth-value. Or, to put it in terms of existence, future states of affairs neither exist nor do not, but have something which we may call an indeterminate existence.

We are back here to the problem about the indeterminacy of the future, postponed from the previous chapter.[5] I postponed it because indeterminacy, in this context, is closely related to problems about time. If the past has determinate existence and the future not, it looks as if we are abandoning the law of excluded middle in a time-dependent way. We are no longer involved with the straightforward metaphysical question about which temporal items exist. Relation to the present time is rather being envisaged as affecting the very concept of existence itself.

The move from the non-existence of the future to its indeterminacy has, as I have pointed out, the advantage of avoiding the unpalatable consequence that all predictions of future events are false. But it also has something important in common with the view it seeks to replace. It continues, along with that view, to reject the simple doctrine of the existence of the future. So any reason people have had for rejecting that view can be claimed in its support as well.

Why then, it may be asked, should anyone be concerned to reject the existence of the future? The picture I have suggested of a growing universe has a positive appeal, but why should it seem so appealing? One central reason, without doubt, is that it is thought to allow scope for freedom of action. If the future exists, it has been supposed, no one can do anything about it, any more than they can do anything about the past. We have here the spectre of fatalism. What will be will be, so there is no room for manoeuvre.

Whereas, on the simple opposed view that the future does not exist, we will, on this line of thought, have ample room for manoeuvre. The trouble however arises that it gives us much more room than we need if we are to avoid fatalism. It makes all free action look as if it were creation out of nothing. It also allows for the application of this notion of creation to the coming into being of natural phenomena like eclipses as well as to things we can bring about like sea-battles. Bergson, who held a theory of this growing universe kind, was prepared to accept this consequence. But most who are concerned with freedom of action will be much happier with much less. This is what they will get if they think of the future existing in a certain sort of way, but in a way which by their actions they can modify. This is where the idea of an indeterminate existence comes in to supply what is needed. The future is no longer a blank, but a kind of raw material to be made determinate by actions. This advantage is no doubt the reason why opposition to the existence of the future has so often taken the form not of simply denying its existence but of claiming it is indeterminate.

I hope it will be clear by now however from the considerations of the previous chapter, that asserting such indeterminacy, and thereby breaking down the dichotomy between existing and not existing, is an excessively high price to pay for the resolution of any problem. This applies to the problems of vagueness and of quantum mechanics, and it applies equally to the problem about the future. We should consider carefully whether there is some other way of dealing with the problem.

Fortunately in this case there is a very direct way, and that is to say that there is no problem and that the appearance of one is the result of confusion. There has been a great deal of discussion of this so-called 'problem of future contingents' since the time of Aristotle, who said that statements about future events were neither true nor false, and that if this were not so, 'there would be no need to deliberate or take trouble'.[6] Without going into the details of this controversy, I will say shortly where I think the confusion lies.

It lies, I suggest, in a failure to distinguish between causal indeterminacy and an alleged indeterminacy of future events themselves. We can speak of causal indeterminacy if, given the present, what happens in the future is determined neither entirely nor in part by causal laws. Suppose we were to admit the absence of universal causal laws, saying, for example, that all regularities were merely statistical, or that the actions of individuals are determined only by their choice, or even that the whole universe is a chaos, everything occurring by absolute chance. We could still despite all this insist that all states of affairs in the world, including future ones, simply existed and were of a determinate kind. The use of the word 'indeterminate' here is perhaps unfortunate, in that it can lead to a running together of causal indeterminacy with what we might call existential indeterminacy, resulting in a confusion between quite different issues.

When it comes to leaving room for deliberating and taking trouble, there are many who have denied that even causal indeterminacy is required for this. But argument on this point need not concern us. Let us suppose that we allow causal indeterminacy in the interests of freedom of action. The question remains why those concerned about freedom of action should not be satisfied with this. Why should not future events, the existence of which is determined by us, be still regarded as existing determinately in the ordinary way?

Given that we are not talking about causal indeterminacy, there seems to me to be no reason at all. When it is said that, if the future exists, we can do nothing about it, what is really meant is that if the future *already* exists, we can do nothing about it. This is certainly true, just as it is true, *pacé* believers in 'backward causation', that we can do nothing about the past. But likewise the condition stated is certainly false—the future, by definition, does not already exist. There is clearly here a trading on an ambiguity in the use of the word 'exist'. There is confusion between 'exist' in its time—independent sense and 'exist' in the special sense of 'exists now'. Because things in the future do not exist now, it does not follow that they do not exist in the time-independent sense. And in fact, when we act with an eye to producing some future state of affairs, we envisage that state of affairs as having a determinate existence, even though it is not yet present. Our only concern is that it should be different in certain respects from what it would have been if we had not acted.

It appears then that any suggestion that the contingency of future events throws doubt on the existence principle is based on a mistake. The problem in this case turns out not to be a problem at all.

TENSED EXISTENCE

We may now turn from the special (alleged) problem about the future to the general one which starts with the idea of the sole reality of the present, and which therefore concerns the past as well. I have pointed out that, if we take the time-independent concept of existence as basic, the theory that only what is present exists is a very difficult one to maintain. I have suggested that perhaps those who have appeared to maintain it have in fact been doing something more. Perhaps, that is to say, they have not intended to offer an alternative metaphysical theory about the contents of the world, but rather to challenge the assumption that the basic concept of existence is time-independent. In particular, the way we use tenses suggests that presentness is part of the concept. And its being part of the concept is presumably what Prior had in mind when he said that the use of the present tense is vacuous and that to exist is to exist now. Whether this resolves any of our difficulties is what we must now investigate.

On the face of it, this will not help if it is taken by itself. In fact, if anything, it makes matters worse. Statements about past and future happenings would be on such a view not so much false as unintelligible, since they would commit us to the presentness of what is not present. However once the time-independent concept of existence is abandoned, the way is opened up for developing the concept further. If existence is being restricted to present existence, why should we not add that there are other kinds of existence corresponding to the other tenses? As well as things which exist, why not include in their distinct respective categories, things which *did* exist and things which *will* exist. The time-reference built in to the concept of existence thus becomes dispersed beyond the present, different time-references giving rise to different kinds of existence.

It might seem that there is a way of avoiding this proliferation of kinds of existence, while still sticking to the idea that basic existence is present existence. For in speaking of 'present existence', it may be said, we are never intending to refer to just that one particular moment when we are, say, thinking our thoughts or putting pen to paper. Every moment that occurs in the whole time-series, we suppose, was or will be present at one time or another. So in addition to the present present, we might say, we have the past present and the future present, the passage of time consisting of the shifting of presentness from one moment to the next.

The recognition of such a plurality of presentnesses has consequences to which we will return. For the present however (i.e. the present present—when this paragraph is being written, or read), we need only note that it does nothing to avoid the need for kinds of existence to correspond to past and future tenses. For we still have to use these tenses when we refer to those things which happen at times other than the present present. We say of them that they *were* present, or *will be* present. So all that is happening is that instead of assigning to them directly some form of past or future existence, we are assigning to them some form of past or future presentness. And since on this view existence is present existence, the tensing of existence comes up in another form. This can be made clear if we adopt a grammatical device suggested by A.N. Prior, whereby we restrict the tenses to the operators 'It was the case that' and 'It will be the case that', and use only the present tense in the subordinate clause.[7] The presentness of existence can in this way be exhibited in combination with an irreducibly tensed reference to past and future. To say of anyone, admittedly, that it was the case that he is eating his breakfast is not idiomatic. We say 'It was the case that he *was* eating his breakfast'. But this can be dismissed as an idiosyncrasy of language, whereby the tense is unnecessarily carried over to the subordinate clause.

The belief in the irreducibly tensed character of existence, by whatever route it is arrived at, has had a wide appeal. It makes possible the avoidance of what many have wished to avoid, the view that everything in the time series, past present and future, simply exists. And it allows for the central, objective reality of the present in that series.

Among those who have adopted this view have been the followers of Wittgenstein. In his Philosophical Investigations, Wittgenstein referred to Augustine's remark that when no one asked him what time was he knew, but when anyone asked him he could not give an account of it.[8] Instead of regarding this as a confession of failure on Augustine's part, Wittgenstein took it as a case where we need only to remind ourselves of what we already know. We use tensed temporal expressions without difficulty, and it is in this that our knowledge of time shows itself. We talk freely about things that have happened, like the extinction of the dinosaurs and of things that will happen, like next year's eclipse, and everyone fully understands what we are saying. To continue to be puzzled about whether past or future exist is to misunderstand the logic of our language.

Following this line of thought, J.N. Findlay in his paper on 'Time: a Treatment of some Puzzles'[9] discusses what Augustine said about the nothingness of the past. If, he says, we are worried about the nothingness of the past stopping us from finding out any facts about the past, we may, if we like, credit it with 'existence, subsistence or any other suitable status'. But if this is going to suggest to us that we could review or revisit the past, we had better go on saying that the past is nothing. In other words we can say what we like about the existence of the past, devising if we like different kinds of existence, like 'subsistence', so long as this does not lead us into conflict with well-understood ways of expressing ourselves about temporal situations.

If it is thought that this Wittgensteinian readiness to admit different tensed forms of existence is due merely to a belief that our language is in order as it is, and to a rejection of all metaphysical speculation as idle, we can turn to Wilfred Sellars' paper on 'Time and World Order'. On Sellars' view, our language is not in order in so far as it contains a timeless present, and we need to impose on it, in the way I have mentioned, a distinction between the timeless use and the present tense use. He then opts for the alternative of treating the timeless use as derivative when we are concerned with temporal facts, and taking it to be equivalent to a disjunction of tensed ones. To say that Eisenhower is (in the timeless sense) a president of the United States is to say that either he *was* president, or *is* president, or *will* be president.[10] Thus the neutral tenseless expressions are not more basic than the tensed ones but presuppose them. The world order, presumably, is to be regarded as one which embodies different tensed forms of existence.

This idea that existence statements are irreducibly tensed has received its most systematic expression in what A.N. Prior has said about tense operators. For him as we have seen not only are all existence statements tensed in being statements of present existence, but they can be prefixed with operators to indicate *when* the things concerned are present. On this basis, he and others have constructed a tense logic in which these operators take their place alongside or as a development from the operators used in other branches of logic.

Such tense logics may be of various kinds, but there is one distinction which it is important for us to make. They may stay on an essentially grammatical level, in which case they may be interesting and useful in systematising ideas about the use of tenses in a language. Or they may be intended to have metaphysical implications, their logic being taken as something which lays bare the structure of the world. It is when tense logic is taken in this second way that it concerns us, for it is then that it has to treat the actual existence of things as having a time reference built into it. It implies that we can no longer ask the straightforward timeless question—what is the case, but only what was the case, what is (presently) the case, what will be the case, and so on for more complex tenses. In doing this, it in fact moves over from having metaphysical implications to having meta-metaphysical ones about different ways in which things can exist.

It will be clear, then, that the idea that existence is irreducibly tensed has been taken up in various ways. What we have to insist on however is that it is unacceptable for anyone who wishes to stick to the existence principle. We may regard it, in other words, as just one more case of hedging when faced with a difficult issue about what exists. We need have no trouble with the grammatical distinction between tenses so long as we regard this as a device for referring to different parts of what exists. But once we start importing the distinction into the concept of existence itself, we land ourselves into a no man's land in which no one knows quite what to say.

In this case, we need to add, we may doubt whether any difficulties we may have in the understanding of temporal sequences are really resolved by such a move. At this point we should return to the implications of the recognition that the present moment shifts and that at one time or another, every moment is the present moment. At first sight, this seems to imply that there is an earlier-to-later time series along which the shift takes place, the shift being from one existing moment to another. But to admit this would be to abandon presentness as a criterion of existence and revert to the simple theory that everything in time exists in the time-independent sense. We therefore need to be more cautious, and speak only of a series of present moments, what exists at each moment, whichever moment it may be, being all that exists. Such a series could then be taken as primitive and as constituting a temporal sequence, rather than being dependent on an existing one. Past and future could then be introduced as having some kind of derivative existence associated with the present in question.

This is the point at which a serious difficulty begins to emerge. This is that the pasts and the futures of different present moments do not appear to be independent of each other. How can we avoid admitting that the very same event can be in the future at one moment and in the past at another? The Battle of Waterloo had a future existence when the present moment was in the eighteenth century and it has a past existence now. How, we must ask, can it be both future and past? If these are really different kinds of existence we do, admittedly, avoid a contradiction.[11] But it is just about as bad to say that one and the same thing can have different kinds of existence. It will be said, of course, it is future in respect of one present moment, and past in respect of another. But this does not help much, since it leaves us with the idea that the kind of existence something has—future, present or past—changes with the shift of the present moment.

Here, I think, we would have reached one of those points where no one would know quite what to say. A natural analogy would be with a change from being a youngest sibling to second youngest when another child is born, or a change in a city from being north of a passing aircraft to south of it, when it flies over. But these are changes in the relational properties of existent things, and to line up the change from future to past with one of these would be to abandon the claim to their having different kinds of existence. If past and future events had such a difference in status, it would be hard to envisage any of them changing its status merely because of a change in its relation to shifting present moments.

TEMPORAL PERSPECTIVES

It looks then as if those who envisage a difference in respect of existence between present, future and past have an internal problem on their hands. There is, however, a modification which might be supposed to avoid the difficulty. What we need is to combine the sameness of an event with a radical change which occurs to it when the present shifts. Might not this be done if the change was thought of as a change in perspective, things viewed from each moment being viewed in a different perspective, sometimes as future and sometimes as past?

What we have said previously about perspectives should make us suspicious of this move.[12] The idea of a perspective, as we have seen, has its origin in the description of relations in space in which one and the same thing can look different from different viewpoints, because of difference of angle and distance. To extend its use, by analogy, to other spheres has been shown to be dangerous, and largely illegitimate. What we need to consider is whether its extension from space to time fares any better than it does in other cases. I think we will find that while it may have a limited point, it does not help with the problem about the difference between past and future existence.

We should start by noting a common misuse of the analogy with space, in discussions about time. It is sometimes said that the theory that everything in time simply exists, in the time-independent sense of 'exist', is one which illegitimately makes of time just one more spatial dimension. The reason for this being said is no doubt that on this theory there is an analogy in respect of existence between time and space. Just as everything in space exists, so, on this theory, everything in time exists. This makes it reasonable to speak of time as a fourth dimension, and moreover to consider its interrelation with the spatial dimensions as envisaged in the physical theory of relativity. But it does nothing to suggest that we are introducing into our account of temporal sequences the spread-out-ness of things in space. The relation of earlier to later remains quite distinct from such relations as that of near to far, though it is often hard to avoid the use of spatial metaphors. Such metaphors may often be harmless, but we should guard against taking them too seriously.

What then should we say of perspectives? Clearly, so long as we stick to the theory that everything in time exists, and that time can be regarded as a kind of fourth dimension albeit not a spatial one, the idea of a temporal perspective, on analogy with the spatial ones, will have a limited use. We can take different 'points' in time as different viewpoints, from which any one occupying them can look back into the past and forward into the future. The perspective here will of course be a one-dimensional perspective, and the analogy will be merely with that of seeing something from different distances, not from different angles. From one viewpoint some events will appear in the more remote past than from another, and there will be some in the near past from one which are in the future from another. Distance or remoteness may lead to error, but this is something as we have seen with which no account of perspectives can cope. Since this applies to the temporal as well as the spatial case, the analogy still holds.

Trouble arises however once we abandon the theory that everything in time exists. And yet it is those who have abandoned this theory and speak in terms of the central reality of the present, who are inclined to appeal to perspectives.[13] One reason that has been given for doing this is that a present moment provides a viewpoint within time itself from which things temporal can be surveyed, whereas if everything in time exists, such an internal viewpoint is lacking and we would need to envisage a 'God's-eye' view of them.[14] But this seems to be based on the mistaken assumption that no series of existing things can be viewed from a point within the series. Let us admit that there could be no 'view from nowhere'[15] as well as admitting that knowledge of temporal sequences has a perspective element. This in itself would be no reason at all for abandoning the theory that everything in time exists, any more than there would be a reason for abandoning the theory that everything in space exists. This leaves us with the question what other reason there could be.

One such seems to be that the appeal to perspectives enables us to avoid the problem we have raised about the difference between past and future existence. If we take past and future as what appear in perspective from the viewpoint of the existent present, should we not expect them to be different for different present moments?

Saying this however does not in fact avoid the problem, though it may give the appearance of doing so. The point remains that it is one and the same thing which is being taken as viewed from the different viewpoints. The Battle of Waterloo may appear in a future perspective for someone in the eighteenth century. Yet surely, if we take the perspective idea at its face value, it remains the very same battle from whichever position it is being viewed. The perspective idea may introduce differences of viewpoint, but its whole point is to reconcile these with the sameness of the object.

One may wonder then how the appeal to perspectives could be of any use to those who think of temporal existence in terms of a series of present moments. It does not solve their problem, though it remains perfectly consistent with the theory that they oppose.

To enable it to solve the problem, there is clearly one thing which needs to be done, and that is to modify the concept of a perspective so as to eliminate the requirement for the sameness of the object. Past and future present themselves as perspective vistas from each present moment. The vistas differ in content, depending on which present moment it is. But we should avoid thinking of them as bound together in any way except through their dependence on their respective present-moment viewpoints. The Battle of Waterloo exists, simpliciter, at the time of its occurrence, but it may be said to exist at other times only in the perspective viewpoint of those times.[16]

We may observe here a parallel with the route along which Karl Mannheim's theory of perspectives developed into the relativism of the sociology of knowledge.[17] No longer do we have different kinds of existence depending on tense, but instead have all temporal existence as relative to the present. The challenge to the existence principle is being shifted from one of its clauses to another. In fact it is being made even more severe, since what we are virtually envisaging is an innumerably large number of different kinds of existence depending on perpetually changing present points of view.

We can only conclude from all this that a theory which gives to existence an intrinsic time-reference is in a bad way. Not only does it mean abandoning the existence principle, but if it is to get out of its own difficulties it has to turn temporal existence into relative existence, with all the further difficulties that that entails. In view of this, it is urgent that we look for alternatives, and, as we have seen, the only acceptable alternative which conforms to the existence principle is one which allows existence, in the simple basic sense, to everything temporal. There would have to be serious objections to this theory if we were to reject it. We have already argued that objections to it based on the alleged indeterminacy of the future are without foundation. Can we say the same of the objection based on the alleged central objective reality of the present?

I think we must confess that the idea of the present time as nothing more than the one one happens to be at, with nothing else to distinguish it from other times, is on the face of it counter-intuitive. Consider the time some thousands of millions of years ago, when the Earth was in a molten state. There was no one to fix that as their present time, no sentient beings were around, and no one to say 'Now'. Yet surely was there not then a present moment extending throughout the whole universe? Considerations such as these need to be taken into account by those concerned to uphold a time-independent concept of existence.

It may of course be said that they are just intuitions and can be explained away as illegitimate extrapolations due to the importance everyone assigns to their present present. But we need to be reminded that there is a possibility of a compromise here of a kind mentioned at the beginning of this chapter. May we not admit a shifting present as an important fact about the world, while still regarding that world as one of existing events or things related to each other in the basic time series of earlier to later? Being present is something which could be said first of one item then the next, and so on, covering the whole series. In this way we could preserve our intuitions without abandoning the simple theory about temporal existence with which we started.

With such an objective present, we would of course also have an objective past and future. But if we start off with an existing earlier-to-later time series, this would not trouble us. Past and future events could be regarded as nothing more than those which are earlier and later than that which are occurring at some present time. And being-earlier-than and being-later-than are relations not properties. The events, being events, do not themselves change (change itself being a relation between events), but they change in their temporal relations to the shifting present.

My point in mentioning this as a possible theory is not to establish it against all possible objections, of which there are no doubt many. It is rather to make it clear that there are alternatives to be discussed within the framework of a time-independent concept of existence. And while these remain, we should certainly reject any theory which moves outside that framework.

NOTES

1 Augustine, *Confessions*, Book XI, Sect. 15, translation as in *Basic Writings of Saint Augustine*, ed. Whitney J. Oates (New York, 1948). Reprinted in R.M. Gale (ed.) *The Philosophy of Time*, (Anchor Books, 1967) p. 44.

2 A.N. Prior, *Time and Tense* (Oxford, 1968) pp. 20-21.

3 Gillian Romney 'Temporal Points of View' in *Proceedings of the Aristotelian Society*, 1977-8, p. 237.

4 Henri Bergson, *Creative Evolution*, trans. Mitchell (Macmillan, 1911) p. 5.

5 See Ch. 6, p. 101.

6 The reference is to Aristotle, 'On Interpretation', Ch. 9. See *The Works of Aristotle*, W.D. Ross (ed.), Vol. I, pp. 18a-19b (Oxford, 1937).

7 A.N. Prior, *Time and Tense*, p. 8 (Oxford, 1968).

8 Wittgenstein, *Philosophical Investigations*, para. 89 (Blackwell, 1958).

9 *Australasian Journal of Psychology and Philosophy*, Vol. XIX (1941) pp. 216 *et. seq* reprinted in R.M. Gale (ed.), *The Philosophy of Time* (Anchor Books, 1967). See esp. p. 155.

10 W. Sellars, 'Time and World Order', in *Minnesota Studies in the Philosophy of Science III* (eds Feigl, Maxwell and Scriven) (Minneapolis, 1962). On p. 533, he makes the suggestion of using 'be' instead of 'is' where no time reference is required.

11 It was McTaggart who claimed that attributions of pastness and futurity to the same thing was a contradiction. He was assuming that the past-present-future sequence consisted of straightforwardly existing items. On this assumption his claim would have point, despite objections. See J.M.E. McTaggart, *The Nature of Existence*, Vol. II (Cambridge, 1927) Book 5, Ch. 33. See also D.H. Mellor, *Real Time* (Cambridge, 1981) Ch. 6.

12 See Ch. 5.

13 See, e.g. Genevieve Lloyd, 'Time and Existence', *Philosophy*, Vol. 53 (1978) pp. 215-228 and Gillian Romney, *op. cit.*

14 See Genevieve Lloyd, *op. cit.*, pp. 215-216. 'There is no question of standing, even in thought outside the flux of time and describing it all without the intrusion of a temporal perspective.'

15 See Ch. 5, p. 78.

16 See Gillian Romney, *op. cit.* For her, temporal viewpoints are intrinsic to time, whereas spatial viewpoints are not intrinsic to space.

17 See Ch. 5.

CHAPTER EIGHT

UNIVERSALS AND PROPERTIES

We must now consider two kinds of alleged entity which might cause trouble for the existence principle. In speaking of what exists, what we normally have in mind is particular things and events. When we say for example that everything in the time-series exists, it is such particular things and events that we have in mind. What now are we to say about the universal kinds of which these particulars are instances. And what about their properties—their intrinsic qualities and the relations between them? Can we say that these also exist? And if so, are they not plausible candidates for having some different kind of existence?

The questions about universals and properties have often been treated as one for the simple reason that properties have been thought of as universal. But it has become increasingly recognised that we should consider particular property instances as distinct, on the one hand, from the universals of which they are instances and, on the other, from the particular things which have them. Hence the question about the existence of properties as distinct from the things should be regarded as different from that about the existence of the universals of which these properties and things are instances. Since, moreover, different considerations are raised by these two questions, it is advisable to hold them apart and at least start by dealing with them separately.

THE EXISTENCE OF UNIVERSALS

The assertion that (in the words of Locke) 'all things that exist [are] particular'[1] has had a long history. But the belief in the existence of universals, which it opposes, has had an even longer one. It was held by Plato, and his 'theory of forms' introduced a realism about universals, which for that reason has often been called Platonism.

On the face of it, this issue about the existence of universals is a straightforward metaphysical one—one about what kinds of thing exist in the world. 'Exist', we would assume, is being used in its simple elementary sense, and in a way which conforms to the existence principle, the metaphysical issue being presented on that assumption.

Why then should this question be of concern to us when we consider the acceptance or rejection of the existence principle itself? The answer is that it has affected this principle in two quite different and opposite ways. Let us take note of these in turn.

(a) Kinds of existence—Platonism

In the first place, if you accept that universals exist as well as particulars, it is very easy to pass over from the view that these are different kinds of existent to the view that they have an existence of different kinds. As two kinds of entity, they are so radically different that it has been found difficult to think of them merely as two parts of the one world. White things, for example, and acts which are just in varying degrees are to be found at many different times and places, but whiteness and justice, taken as universal properties, have no particular temporal or spatial location. These unlocated entities clearly do not exist 'alongside' the spatio-temporal items which are their instances. Hence the inclination to envisage them as belonging to 'another realm' with an existence of a different kind.

The inclination to think in this way is one which affects any radical metaphysical dualism. I have suggested that Descartes did not succumb to this, and that he regarded his radical dualism of mind and matter as a straightforward metaphysical one, thus allowing for causal interaction between his two kinds of substances. But it is noteworthy that his successors thought it necessary to disallow any such interaction between the spatial and the non-spatial, and to assign to them different kinds of existence in two different worlds. When we come to objects which are not only spaceless but timeless as universals are supposed to be, it becomes even more difficult to avoid regarding them as belonging to a world quite different from that of spatio-temporal things.

It would be hardly surprising then if Plato, who was concerned to insist on the reality of forms, should have succumbed to the temptation, as is commonly supposed, to assign to them a reality, or a degree or kind of reality, which other things did not possess. As I have mentioned,[2] he appeared, specifically in a notable passage in Book V of the *Republic*, to challenge the law of excluded middle by representing the 'world of sights and sounds' as hovering somewhere between full existence and non-existence?[3] His use of the word 'full',[4] as I have said, suggests that he may have had in mind not so much a challenge to this law, as an assignment to the world of sights and sounds of a lesser degree of reality or a different kind of reality. Whichever way we take it, the important point is that he seemed prepared in one way or another, in the interest of his theory about universal forms, to override the existence principle.

One way, as I have pointed out, in which to give some pictorial content to some third alternative between existing and not existing, or to some lesser degree or kind of existence, is to use the metaphor of shadows.[5] And I have referred to the fact that this is something which Plato did in a vivid way in his story about the cave, thereby committing himself to the idea of some special existence (or non-existence) status for the sensory world. And while Plato himself made no explicit appeal to another way of picturing a third alternative—that of treating the sensory world as one of appearances, the reference to appearances has been made much of by commentators in their account of his views.[6] So once again the reality of universals and the rejection of the existence principle are seen as going together.[7]

However, despite this tradition starting with Plato, the temptation to adopt a 'two worlds' solution is one which has to be resisted. In the case of particulars and universals there is in fact a specially strong reason for resisting it. This is that it seems impossible for anyone adopting this solution to give up the idea that the two worlds are positively related by the relation of instantiation. Descartes' followers may have been prepared, in the interests of an existential dualism, to give up all relations between the mental and the material. But no believer in the existence of universals in one world can avoid admitting that the particular things in the other world are their instances. And such a relation between entities in different worlds is, I have argued, impossible.[8] Instantiation therefore would remain a mystery.

It might be thought that it would make some difference here if we denied that there were uninstantiated universals. Plato clearly envisaged universals which had no instances when he included among them unrealised standards of perfection, and this, it might be said, was what put him on to the idea that they occupied an entirely separate realm. If the separation of realms is to be resisted, therefore, it might appear that all we need to do is to abandon uninstantiated universals.

This however would be a mistake. It must be recognised that the postulation of 'separate realms' or 'two worlds' may mean two quite different things. It may mean merely that we are envisaging two parts of the one existing world which are in large part out of relation to one another. Or it may mean that we are taking them to be distinct in respect of their existence. Admission of uninstantiated universals, clearly, need mean only the first.[9] We could admit a vast array of them—one, perhaps, for every meaningful expression in a language except for proper names—and still regard them as a very extensive part of one world, though unrelated by the relation of instantiation to the other part. This being so, abandoning them would make no difference. The special problem with the 'two worlds' view dealing in different kinds of existence is not its undue proliferation of universals which are not instantiated, since this could take place without it. The problem is rather that of making sense of the instantiation of those that are instantiated.

Assuming then that we reject all idea of there being degrees or kinds of existence, we need to consider what other ways there are of dealing with the problem. Possible ways of dealing with it must clearly be one or the other of two kinds. One is to reject the metaphysical dualism, thereby removing any motive for turning to a meta-metaphysical two worlds doctrine. The other is to accept some form of the dualism and then brazen it out, denying that it requires us to play around in any way with the elementary concept of existence. It is not our business to decide the metaphysical issue between any of the theories of these kinds. But something should be said about them to indicate how the metaphysical issue might be intelligibly discussed in these terms.[10]

The commonest monistic solution is of course the denial, as by Locke, of the existence of universals. This has often been called nominalism, but it is not to be restricted to the view which identifies universals with particular names, any particulars or sets of particulars will do. One of its main forms has been that in which resemblance between particulars has been taken as the basic notion, instead of them all being regarded as instantiations of one single universal entity. Though this is the commonest solution, we should keep in mind its converse, which has denied the existence not of universals but of particulars, these being identified with collections of universals. Whatever their relative merits, such solutions have the obvious advantage of simplifying the theory, and removing at one stroke all ground for thinking in terms of two levels of reality.

We may well ask, however, whether such reduction is necessary, in whichever direction it goes. Why not simply allow that these two very different kinds of entity exist as parts of the totality of things, and that members of the one are instances of members of the other, all within the comprehension of one world. Admittedly, the one lot—the universals—are not spatio-temporally located, and the others are, at least for the most part,[11] commonly supposed to be, but, it might be said, spatio-temporal location is one thing and existence quite another, there being nothing in the concept of existence to prevent the inclusion both of what is spatio-temporal and what is not.

I see no reason why we should not in this way treat the dualism of universals and particulars on its merits. The issue then comes to be what merits it has. Though it passes our existence test we could still have serious doubts about it as a straightforward metaphysical dualistic theory. Though we may now admit instantiation as a relation between different parts of the one world, it remains a relation of a very peculiar kind. It is a one-many relation between an unlocated, timeless entity and an indefinitely large number of entities in particular times and places. Any relation of the eternal to the temporal, and of the spaceless to the spatial, will always be a problem, and anything that avoided it would be better.

(b) Dependent existence—Aristotelianism

This is the point at which we come to the proposal, dating probably from Aristotle, that universals should be 'brought down to earth' and conceived of as existing somehow within the spatio-temporal world. This would automatically eliminate the uninstantiated and remove the problem of the relation of the eternal to the temporal. There would be no separate realm in any sense of this phrase.

It must be admitted that this would not entirely solve the problem about instantiation. Though we would no longer have to envisage some relation between the eternal and the temporal, we would have to think of the universal, being 'in' each of many particulars, as being multiply located. It may be said that we should not be worried about this. Universals are after all not particulars, which *are* spatio-temporally located, so we must apply different principles, and grin and bear it.

There is however more serious trouble in store. For it is here that we come to the second of the two ways in which views about universals affect the existence principle. When the universals are 'brought down to earth', there is no longer a question of there being different kinds of existence belonging to different worlds. But another question about kinds of existence emerges in its place. It becomes necessary on this view to deny that universals can exist in independence of the particular instances with which they are associated. And if we ask what kind of dependence is involved here, we can find no other than an existential dependence, of the same kind that Berkeley had to admit for his ideas. All the problems involved in this concept of dependent existence come once again upon us.

It is true that the context in which these problems now arise is very different. The issue is now about the dependence of universals on particulars, not of ideas on minds. But the argument about dependent existence remains the same. We start with a statement of a necessary condition. If there is no particular instance there is no universal. We then look for some ground for this assertion. The connection is clearly not a causal one. Nor is it one of logical necessity; there is nothing in the concept of a universal, even when it is given some spatio-temporal multi-location, that entails its having particular instances. There seems no alternative but to say that its existence only in connection with particulars is part of its nature, that its dependence on particulars in other words is descriptive of the kind of existence it has. We may speak of this dependence as categorial if we like, thereby dissociating it both from the causal and the logical, but this is virtually another way of saying the same thing.

I have presented this situation as if it would only come about with the bringing of universals down to earth. But we could here point out that any theory about universals which disallowed uninstantiated ones would imply the same kind of dependence relationship. For we need to give some reason for denying existence to the uninstantiated ones, and what reason could we give except that they were dependent for their existence on their having instances, this dependence as before being neither causal nor logical? It is the intrinsic nature of this dependence which causes trouble and has to be resisted, in connection with whatever kind of theory of universals it arises. Though there is no doubt that giving universals a spatio-temporal location 'in' the particulars is what brings it to a head.

Despite the parallel, there is one big difference between mind-dependence and particular-dependence which I have not mentioned. This is that in the case of particulars and universals the dependence has to be regarded as reciprocal. If universals are dependent for their existence on particulars, particulars are also dependent for their existence on universals. The alternative is that we envisage the possible existence of 'bare particulars' which are not instances of any kinds. And these seem even a less reasonable thing to postulate as part of the world than the uninstantiated universals to which they correspond.

If we accept this reciprocity of dependence it is clearly going to make matters worse. We cannot any longer think of anything as simply existing (in our elementary sense of 'existence') as we could, for example, with Berkeley's spirits. There are two kinds of dependent existence, and that is all there is. The position becomes more like that envisaged by believers in relative existence who dissolve the world into a multiplicity of points of view. But because we have a mutuality of dependence, there are in this case no external anchor points—like points of view. There are just existence-dependent-on-particulars, and existence-dependent-on-universals, and that is all.

In presenting this problem about universals, I have made no mention of properties. But since a similar problem arises about properties and the things which possess them, we should introduce these, before considering what can be done about the situation in both cases.

THE EXISTENCE OF PROPERTIES

The trouble is that, even if we were to accept the monistic theory that everything which exists is particular, we would still have two completely different kinds of particular on our hands. On the one hand there would be the things and events of which we ordinarily take the world to be composed and on the other there would be the properties and relations which we ordinarily suppose they have. These properties and relations would be the particular ones belonging to particular things and events, so no problem about the existence of universals would arise. Yet we would still be left with the dualism of things and their properties. I leave out relations and events here, for simplicity's sake, though something will have to be said about them later. For the present let us talk only of things and their properties.

This dualism, we need to insist, is to be distinguished sharply from that of particulars and universals, despite the traditional use of the word 'property' to refer to universal properties of which particular properties have been called 'instances'. To avoid confusion, we might drop the word 'property' and refer to these particular property instances as abstract particulars, or adopt a recent practice of using a different word for them like 'trope'.[12] The dualism we are here to consider would then become the dualism of concrete particulars and abstract particulars, or tropes.

This dualism might be regarded, like that of particulars and universals, as a straightforward metaphysical one. But it has been found easy, as in the case of particulars and universals, to think of it as a dualism which does not so much presuppose the elementary concept of existence as throw doubt on it. It might do this in two different ways which run parallel to those discussed in the universals case. Firstly, the great disparity in kind between things and properties might lead to their being thought of as having two different kinds of existence. And secondly, once it is seen that properties could not exist without things they come to be thought of not simply as existing but as having an intrinsically dependent existence.

On the move from great disparity of kind to there being different kinds of existence, we should realise first that there is little to be gained from analogy with the case of particulars and universals. There has certainly been no Plato to uphold the greater reality of properties as there was to uphold the greater reality of forms. As far as the grading of reality is concerned, the move has in fact been the other way. Things, spoken of as substances, have been thought of as the real stuff of the world, their properties having only a subsidiary kind of existence.

The disparity nevertheless remains, and the question remains what it justifies in the way of a challenge to the existence principle. Since we are now dealing only with particulars, it must be pointed out that the main persuasive consideration which made for a separate world of universals is missing. Particular properties belong to particular things at particular times and places, and furthermore they are not multiply located. This makes it hard to think of them as inhabiting a separate world. Given that both substances and properties exist, there is of course a great disparity between them. But no such disparity in kind ever entails a difference in kind of existence, so in the absence of very special reasons, we should not take the attempt to move from the one to the other too seriously.

But this is where we come back to the second kind of challenge, and this cannot be dismissed so easily. If we are to have both things and properties, it seems clear that the properties are dependent on things for their existence. There are, as it were, no free floating properties which are not the properties of anything. And this dependence like the dependence of universals on particulars, is neither causal nor logical, it is simply part of the nature of the dependent entity. So we have a problem about dependent existence similar to the one we have when we have 'down-to-earth' universals.

It might appear, admittedly, that in this case there is a difference, in that the dependence is not mutual. The things which form one party to the dualism, are often spoken of, as I have mentioned, as 'substances', and the dualism as a dualism of substance and property. And it is commonly taken to be part of the meaning of 'substance' that it refers to what stands on its own and needs nothing else in order to exist. It might seem then that we have in this case, not two entities each with its own kind of dependent existence, but rather one as simply existing and the other as dependently existing.

This however would be a mistake due to confusion here about what counts as a substance. The etymology of the word does, admittedly, suggest something which exists in its own right, and could exist in the absence of any properties. For it offers us the metaphor of that which 'stands under', or 'supports' properties and what is supported can be removed without removing the support. The use of the word 'substratum' as by Locke[13] is designed to emphasise the same idea, by presenting us with a non-reciprocal dependence of the top layers on the bottom layer.

But these, of course, are metaphors, and they can prove deceptive in this context. It is not merely that all descriptions of entities make reference to their properties so that a substratum becomes, as Locke admitted 'something we know not what'.[14] It is also, as Berkeley insisted,[15] that the whole notion of 'support' is in this context empty, and merely serves to repeat the assertion that there is some part of an ordinary thing which can exist without its properties. This is something we have no reason to suppose. A propertyless substratum suffers the same fate as the bare particular. There is, of course, an asymmetry in the contrast between things and properties, in that one thing has many properties, and not vice versa. When we call things substances, this no doubt is part of what we have in mind. But this does not stop us from saying that things depend on some set of properties or other for their existence just as properties depend on things. And this dependence is certainly not causal, in the way the metaphor of support suggests.

We have then here a genuine problem, not only about dependence but about mutual dependence, of the same kind as we have about the universal-particular connection. As in that case, it would be a problem that would disappear if the dualism was dropped and it was said that what existed was only things, or only properties. Both these kinds of monistic theory have been held. On the one hand there is the theory that having properties is no more than being members of classes, which after all, are collections of things.[16] On the other hand it has been maintained that things are collections of properties.[17] However, such reductive theories, though they would run into no trouble as far as the existence principle was concerned, do run into considerable difficulties as straight metaphysical theories. It would therefore be highly desirable if the problem could be solved in some other way. How this might be done is something we must now consider having in mind both the case of universals and the case of properties.

EXISTENCE AND MUTUAL DEPENDENCE

One thing we might do of course is simply to deny that the dependence in question involves us in admitting that the dependent entity has an existence of a special kind. Why not, it may be said, simply accept particulars and universals, things and properties, as existing in an ordinary unqualified way, and regard the dependence in question as a special kind of relation between them. We admit different kinds of dependence relation between existents, such as the causal and the logical. Why not include existential dependence, as I have called it, as just another kind of which a constructor of metaphysical theories has to take account. In that way we would avoid any breach of the existence principle.

It should be noted that this proposal is different from that adopted by realists in opposing the idealist theory of the knowledge-dependence of ideas. In that case the realist solution was to deny that there was any dependence relation at all; objects of knowledge could exist apart from the relation of being known, even if they did not actually do so. What is being proposed here is different in that it admits the dependence but denies that it affects our notion of existence. And it points to the cases of causal and logical dependence as being precisely of this kind.

We should observe that when it is said with respect to any alleged relation of dependence that the objects related exist in their own right, there can be some uncertainty about which of these two positions is being taken. One thing which may be meant is that there is no such relation of dependence, and that the allegedly dependent object can exist in the absence of the other. The extreme position of this kind would be that there were no dependence relations of any kind and that any object could exist in the world in the absence of all others. Hume, for example, has had this extreme position attributed to him though he nowhere states it. What he denied was that there were relations of logical dependence. 'There is no object', he said, 'which implies the existence of any other, if we consider those objects in themselves.'[18] Whether he also denied causal dependence or dependence of any other kind is a matter of dispute. But this is not an issue which need concern us. We are concerned with whether dependent objects 'exist in their own right' in the weaker sense of their dependence not affecting their existence. Let us admit that objects which are causally, and even logically, dependent on others exist in their own right in this sense. What we are concerned with is whether what is acceptable in the cases of causal and logical dependence can be extended to other cases of a dependence relationship.

The general question raised here can best be left to the next chapter, in which we consider the notion of necessary existence. So far necessity and its related possibility have not been mentioned, and they are concepts which need close scrutiny. We can see at once however how a dependence relationship might be treated quite naturally as a necessary one. To say that one thing is dependent on another is to say that it could not exist without it, not merely that it does not do so. And put another way, this is to say that that other thing is necessary for its existence, or to use another phrase is essential for it. In discussing dependent existence, we seem in fact to be discussing a special case of necessary existence, and the issue here raised can be considered later as a general one in those terms.

Here however let us restrict ourselves to the even more special case of the mutual dependencies of substances and properties and of particulars and universals. The proposal is that these should be regarded as straightforwardly existing objects which depend on one another in the same kind of way in which effects are dependent on their causes, or what is entailed is dependent on what entails it.

Though this would solve our problem, we have to realise that the analogy on which it rests is not a good one. It is not merely that things and properties are items of very disparate kinds in a way that items causally or logically related are not. Nor is it merely that properties are not the sort of items, as distinct from things, that we commonly speak of as existing. It is rather that to classify some dependence as causal or as logical is to provide it with a context which is lacking in the case of the dependence of property on thing, or universal on particular.

It will be remembered that in discussing the alleged mind-dependence of ideas,[19] I said that the mere statement that if one item did not exist the other would not do so is one that needs some ground if it is to be acceptable. Appeals to causation and to logic provide such a ground. Causal and logical relations are kinds of relations which we can say a lot about quite apart from their presence in some particular cases. But the alleged substance-property dependence, like mind dependence, has to stand on its own without anything more to be said. That is why, in both cases, we are inevitably led to account for it by appealing to the intrinsically dependent existence of the items. And what is said here about things and properties applies equally well to the mutual-dependence relationship between particulars and universals.

How then can we avoid this unacceptable conclusion? There is one other way, which may be said to be a fairly radical way, but which is well worth following up. Taking the case of things and their properties, this is simply to deny the existence of both of these alleged items and to say that what exists is propertied things and propertied things only. We have after all already denied that there are things without properties or properties without things, and this is not just because there are in fact none of these to be found. It is because the items are assumed to be connected by some kind of existence-dependent necessity. So all we need do is replace the dependent existence of the items with the simple existence of the state of affairs in which they are embodied. And likewise with particulars and universals, on the theory that neither can be reduced to the other. If we deny that there could be bare particulars or uninstantiated universals, we may say that this is because neither particulars nor universals exist at all, but only states of affairs in which the elements of particularity and universality are incorporated.

Pointers to this solution are not new. In his 'Tractatus' Wittgenstein announced that 'The world is the totality of facts not of things (der Tatsachen, nicht der Dinge)'. In the same vein John Anderson insisted that the world was propositional in character. Others have spoken of 'states of affairs', as I have just done. David Armstrong suggests that we should think of a world as a world of states of affairs, with particulars and universals only having existence within them.[20] I call these only pointers to the solution because they stop short of explicitly denying the existence of the constituent items. Thus when Wittgenstein goes on to speak of things he says only that it is essential to them that they can be constituents of facts. Armstrong says they have existence, but only within states of affairs. In similar vein, C.B. Martin has said, in defending the views of Locke, that properties and substrata are not objects, but ingredients in objects, without saying whether these ingredients exist.[21] Such statements, in fact, retain the idea that these constituents or ingredients have some kind of dependent ontological status. They go little beyond what was said by Edmund Husserl, who never thought of challenging the status of properties as dependent existents, taking them as 'moments' which required a foundation from which they were inseparable.[22]

It will be clear then that in order to solve our problem on these lines, we would need to go further than these writers have done and explicitly deny any kind of existence to these allegedly dependent entities. We would need to stick to the sole existence of full-blooded propertied things, or states of affairs.

When this is done, we should realise, we have virtually abandoned the dualism with which we started. There are no longer two kinds of existent—things and properties, particulars and universals, but only one—facts, propositions or states of affairs. This is once again a monistic solution, but one of a less reductive kind. That we would need to resort to such a monism is hardly surprising. Once we have abandoned hope of interpreting the situation as one of a relation holding between two simply existing items, this is clearly all that can be done.

We should not be blind to the fact that this solution has problems of its own. For one thing, if we are to think of thing-plus-property as a single fact or state of affairs, we have to keep in mind that any given thing has many properties, the totality of which we might call its 'nature'. Are we to deny existence to anything short of the thing with all its properties, or are we to allow that we have an existing fact in the case of each of the thing's properties?

So long as there is no mutual dependence among properties, it will clearly be sufficient for our purpose to admit as a fact a thing with any one of its properties. If a thing (say a ball) is both red and round, we can easily admit that there are here two states of affairs (among many others)—its being red and its being round, since these states of affairs are not dependent on one another. We must also of course allow as a state of affairs its being both red and round, and so on up to that state of affairs in which we include its total nature. We might draw an analogy here with the distinction between parts and wholes. So long as we are prepared to accept that both parts and wholes exist,[23] we can allow existence equally to ever smaller or ever larger parts right up to the whole. Similarly we may allow existence equally to ever narrower or ever more comprehensive states of affairs. There is a difference of course that we have to admit, and that is that all the items in the state-of-affairs series have a constituent in common, the particularity of the one thing which has the properties. If we are prepared to accept that, so far so good.

Having a constituent in common is something we also have to be prepared to accept if properties are taken to be universals. If the ball and the apple are both red, there are clearly two states of affairs, but we have to admit that on such a view the property of redness is to be found in both, just as we have to admit that the particular ball features in both the state of affairs of its being red and the state of affairs of its being round. If this sounds odd in either of these cases, it may be said that this is because we are continuing to think of the things and the properties as entities, and not, as on this view we must do, as no more than distinguishable elements in states of affairs.

A problem would remain however if there were any mutual dependence among properties. And as we all know there are cases where it looks as if there must be. If the ball has shape it must also have a size. If it is red it can't be green. And so on. Does this not mean that the state of affairs of its having a shape is dependent for its existence as a state of affairs on that of its having a size? And so on, in reverse in the case of exclusion of properties.

Whether these are genuine cases of existential dependence as distinct from a merely logical one is something I will not discuss here.[24] But if they cannot be disposed of in this way, what we have to do is simply deny that the ball's having a size and having a shape exist as distinct state of affairs. We must admit, in other words, that there are states of affairs in which the property element is complex, a case in point being that of being a solid spatial object, which as a single property includes within it the having of both shape and size. Similarly, having a uniformly red surface includes within it its not being at the same time of any other colour. In this way the existence—that is the undifferentiated existence—of all states of affairs might be defended.

This is the point at which something must be said about alleged determinable properties. After the considerations about determinacy brought out in Chapter 6, we clearly need to insist that all existing states of affairs are determinate. The ball is of a quite specific shade of red, and we cannot add to this as further facts about the world that it is red, say, and that it is coloured. The same is true about shape and size, and it should be noted that in this case, it is only at the determinable level that the trouble about existential dependence arises. This does not mean that we can ignore the determinable level, for after all red things *are* coloured and round balls *have* a shape. What we have to do is to proceed as before, and regard all determinables as included as elements within the actual states of affairs of which the world is composed.

There are troubles enough here, but their suggested solution obviously leads on to a final serious one. What are we to say about all those constituents or elements which we are constantly finding within state of affairs? We do distinguish and talk separately about things and properties, about different elements within complex properties, about determinables as distinct from determinates. How can we do this without assuming their existence?

Faced with this problem, some have made attempts to deny that we ever do distinguish such elements. Notable among these is David Hume. Hume maintained, in a well-known passage, that what is distinguishable is separable, basing on this principle his denial that there could be abstract ideas. By this he presumably did not mean that the distinguishable items were independent, though as we have seen he denied their logical dependence. What he seems to have had in mind is that they were separable in respect of their existence, for he goes on to ask how in view of this we can account for the 'distinction of reason' we make between (for example) a body and its figure. To which he answers that we do not distinguish them, but only view them in different aspects as we compare them with other things.[25]

Without going any further into the argument of Hume's day about abstract ideas, we can see that this attempt to deal with the problem will not do. If we admit that a thing can be viewed in different aspects, how can we avoid distinguishing these aspects? It seems that the only thing we can possibly do is admit the 'distinctions of reason' but deny that what is distinguished exists.

Hume's reference to 'aspects' suggests that this second way of meeting the objection is not as unreasonable as it might seem. The problem in fact is similar to that which had to be dealt with in the case of appearances.[26] Appearances, like the distinguishable elements in states of affairs, may well be said not to exist yet, like them, they have to be distinguished as intentional objects. The problem is what to do with such 'third items' in an existing world. Any way we found of solving that problem might well be used in the present case.

One way of solving it for example was to deny the existence of these items in the non-mental world, but to admit their existence as mental representations. Similarly we might say that a distinction of reason does after all require the existence of something, but not necessarily of anything apart from us who make the distinction. We do envisage separately things and their properties, even common properties, the shapes and sizes of spatial objects, the determinables under which we classify determinate properties, just as we envisage the various appearances of things. But this, we may argue, says nothing about their existence in the world that we envisage.

Whether we accept this 'representationalist' solution or not, we should keep in mind that the analogy with the appearance problem eases our path when we seek to drive the necessary wedge between the conceptually distinguishable and the existent.[27]

RELATIONS AND EVENTS

We must now, finally, reintroduce relations and events, and make sure that they cause us no further troubles than those that arise in the case of properties and things.

Relations, as we know, are often treated as a kind of property—ones which are dyadic, triadic, and so on, instead of simply monadic, the supposed difference being merely that they involve two or more terms instead of only one. But to treat them in this way is to neglect a further important difference. Things that are related are already things with properties, that is, 'monadic' properties, so that such properties are presupposed in any talk of relations. If relations existed, their existence would be, admittedly, a dependent existence like that of properties, but it would be a dependence at a new level.

In view of this, there is a way of avoiding the admission of such a dependent existence of relations, at least in many cases, which is not open to us when we deal with simple properties. These are the cases in which, once given the properties of the things related, the relation is already given and nothing more is needed. If one ball is 3 inches in diameter and another 2 inches in diameter then one simply is larger than the other, and it does not appear that in asserting this relation between them we are adding anything to our account of the world. We lose nothing therefore if we simply deny its existence.

Hume, it may be recalled, divided relations into two classes—into 'such as depend entirely on the ideas', and 'such as may be changed without any change in the ideas'.[28] If we read 'states of affairs' for 'ideas', the cases I have mentioned are those that belong to the first of these classes. In this he included all cases of resemblance, degrees in quality, and proportions in quantity and number. We could include logical relations. If this class could be extended to include all relations, our problem could be solved by denying the existence of relations in general over and above the propertied things related.

Unfortunately this does not easily fit the cases of spatio-temporal and causal relations, which Hume understandably puts into his other class. If two balls are 2 feet away from one another, there would appear to be nothing in the nature of either ball from which this would follow. And unless we assume some kind of hidden causal power, the same could be said when one ball hits another. It would be as well, therefore, if we had another way of dealing with these cases.[29]

That there is another way will be clear if we consider the possibility of extending to relations what has been said about states of affairs. We may deny that relations exist, yet admit that they are distinguishable elements in relational states of affairs. The things related do of course have properties, and we have allowed that each of them having any given property is itself a state of affairs. What we need now to allow is that both of them together with the relation between them constitute another state of affairs. In this the thinghood of each along with the relation would remain as distinguishable elements. If we do not like the thinghood of a thing being common to different states of affairs, or if we do not like distinguishable elements being said to have no existence in themselves, our position is not much worse than when we were only dealing with properties. All that we have done is to increase substantially the range of what is to count as an existing state of affairs.

Adding events to things may be thought to involve us in a problem similar to that of adding relations to properties. For though we normally assume that events exist (what else do we mean when we say they occur?), their existence appears to be entirely dependent, in some non-causal non-logical way, on the existence of things. There are handshakes, atomic collisions, earthquakes, and revolutions, but there would be no such happenings apart from the people and the things which participate in them.

As in other cases, one might try a monistic solution to this problem, saying either that events can be eliminated from an account of the world or that the world consists of them. But fortunately there is also a way out, similar to the second of those suggested for relations, which can be effected by a further modest expansion of the view of the world as consisting of states of affairs. In referring to states of affairs so far we have made no mention of change, and talking of a 'state' even seems to preclude it. So what we need to do is to point out that things not only have properties, but in the course of their endurance through time, change their properties, and moreover in the course of time begin to exist and cease to exist. Events then may be said to be simply that sub-class of states of affairs which consists of things changing at given times. Often these will include many things and the relations between them. And any event, however simple or however complex, may as a state of affairs be said itself to have properties.[30] With such an account of events, if it can be sustained, we would have no problem about accepting that they exist in the same straightforward way as other states of affairs.

I have said enough to suggest ways in which this theory of states of affairs might be maintained without a breach of the existence principle. If its troubles are thought too great there is always the alternative of resorting to one of the more reductive forms of monism. We must remember however that these too have their troubles. The important thing is that the search for a solution should go on without our getting stuck with such unacceptable notions as that of there being different kinds of existence.

NOTES

1 John Locke, *Essay concerning Human Understanding* (ed. P.H. Nidditch, Oxford, 1975) Bk. III, Ch. 3. Opening sentence. In fact he introduced this not as a separate assertion but as an opening clause, treating it as a matter of course.

2 See Ch. 1, pp. 5, 11.

3 Plato, *Republic*, Book V, M.P. 479 (trans. Lee) (Penguin). The Greek 'einai', it should be noted, is translated differently by different writers. While Lee uses 'existence' Cornford uses 'reality', Jowett 'being' and Shorey 'that which is'. This indicates the ease with which these words are all used to refer to existence in what I have called the basic sense.

4 Other translations are 'pure' and 'absolute'.

5 See Ch. 6, pp. 99-100.

6 See, e.g. R.C. Cross & A.D. Woozley, *Plato's Republic* (London [Macmillan], 1964) Ch. 7, pp. 139 *et seq.*

7 That Plato himself was committed to their being kinds of existence has been disputed. See, e.g. Gregory Vlastos, 'Degrees of Reality in Plato', in his *Platonic Studies* (Princeton, 1973). Also Julia Annas, *An Introduction to Plato's Republic* (Oxford, 1981) Ch. 8. But it remains as the traditional and commonly accepted interpretation.

8 See Ch. 2, p. 26.

9 D.M. Armstrong is one who connects uninstantiated universals (which he rejects) with a 'two worlds' view, but not in the sense in which it requires different kinds of existence. His rejection of them therefore has to be on other grounds. See *Universals*, Ch. 5, Sect. I, esp. pp. 76-77 (Westview, 1989).

10 See Armstrong, *op. cit.* for a useful classification of such theories.

11 God, for example, might be regarded as a non-spatio-temporal, yet particular being.

12 D.C. Williams was one who used this word for them in his *Principles of Empirical Realism* (Illinois, 1966). It is adopted by Keith Campbell in his *Abstract Particulars* (Oxford, 1990).

13 See John Locke, *Essay concerning Human Understanding*, Book II, Ch. 23, Sect. 1.

14 Locke, *op.cit.* II, 23, Sect. 2.

15 Berkeley, *Principles of Human Knowledge*, Sect. 16.

16 See e.g. Anthony Quinton, 'Properties and Classes', in *Proceedings of the Aristotelian Society*, 1957.

17 See e.g. D.C. Williams, 'The Elements of Being', in *The Principles of Empirical Realism*, 1966. Also Keith Campbell, *Abstract Particulars*, 1990.

18 David Hume, *Treatise of Human Nature*, Vol. I, Part 3, Sect. VI, first paragraph.

19 Ch. 3, pp. 30-31.

20 Wittgenstein, *Tractatus Logico-Philosophicus*, Prop. 1.1. John Anderson, *Studies in Empirical Philosophy*, p. 4 and throughout. David Armstrong, *Universals*, Ch. 5, Sect. 5.

21 C.B. Martin, 'Substance Substantiated', in *Australasian Journal of Philosophy*, 1980, pp. 7-8.

22 E. Husserl, *Logical Investigations*, III, Sects 2-5, 17, 21 (trans. J.N. Findlay (London, 1970) Vol. II, pp. 435 *et seq*). For Husserl, properties, as 'moments', and inseparable parts of wholes, are distinguished from pieces, or portions of wholes, which are separable. Note that when Armstrong (*Universals*, pp. 91-93) distinguishes properties from parts, he is treating parts as separable (Husserl's portions) and properties as not (being only constituents of states of affairs). On the idea of essential parts, see also Peter Simons, *Parts, A Study in Ontology* (Oxford, 1987) where Husserl's distinctions are developed more elaborately.

23 I ignore the view that wholes are themselves dependent existents, being made up of their parts. This view would involve us in an atomism which held that only things without parts exist. But if atoms can exist without parts, why not you and me, since after all we might have been indivisible? I am

assuming, that is to say, that existing and being divisible are concepts with no necessary relation to each other.

24 We need only recall the long-standing debate about necessary synthetic propositions in which such illustrations figured largely.

25 *Treatise of Human Nature*, Vol. 1, Part 1, Sect. 6. He actually moves the illustration from body and figure to figure and colour, but in each case the same point is presumably intended.

26 See Ch. 4.

27 Note that J.P. Moreland, in criticising K. Campbell's theory of tropes (*Australasian Journal of Philosophy*, Vol. 67 (1989) pp. 379-393), denies that this wedge should be driven. He does this because following Husserl he sees no problem about dependent existence. If there were no such problem, ways of avoiding it would of course not concern us.

28 Hume, *Treatise of Human Nature*, Part III, Sect. I, first paragraph.

29 See Keith Campbell, *Abstract Particulars*, Ch. 5, for an argument that all relations can be dealt with in the first way. He maintains this by treating position in space and causal powers as properties.

30 For a view of this kind, see for example Jaegwon Kim, 'Events as property exemplifications', in *Action Theory*, ed. Myles and Walton (Reidel, 1976) pp. 159-177, see also Barry Taylor, *Modes of Occurrence* (Blackwell, 1985). Both these writers are concerned to insist, against Donald Davidson, that such a view is consistent with the point he makes that events have various descriptions.

CHAPTER NINE

NECESSARY AND POSSIBLE EXISTENCE

The idea that some things exist necessarily and some only contingently is an old one. Necessary existence has been attributed in particular to God. But we also say of much more mundane things and states of affairs that they must exist or must be the case. In a parallel way we may say of them that they could exist or be the case. The 'must' and the 'could' are the signs that we are speaking of what is necessary and what is possible. We need to consider whether this means that there are after all different kinds of existence, the necessary and the possible as well as the contingent and the actual.

In discussing the mutual dependence of things and properties, I pointed out that this could be taken as a case of something which is necessary. For any property we say there must be some thing or other which has it, and any thing must have some set of properties or other. I contrasted this with cases of causal and logical dependence, which can similarly be taken as necessary. But in mentioning these as cases of necessity, I said nothing about what this necessity consists in, or about how the necessary is to be distinguished from the accidental or the contingent. The easiest thing, no doubt, is to give illustrations; we are after all capable of using words like 'must' and 'can' effectively. But if we are to say anything about the legitimacy of speaking of necessary and possible existence as kinds of existence, we would like to be able to say something about what these illustrations illustrate.[1]

NECESSITY AS A RELATION

A first step in the characterising of necessity is to point out that in all the normal cases in which we make reference to it, it is to be found in the relation between two or more existing items, the terms of the relation. A ball released in a vacuum will of necessity fall 16 feet in the first second. It being the case that Socrates is a human and that humans do not live forever, it is necessarily the case that Socrates will not live forever. Given a perfect triangle, its angles are necessarily equal to two right angles. If you add five balls to seven in a bag there will necessarily be twelve in it. If a letter-box is red, it must be coloured. Let us classify these cases as causal, logical and mathematical, with a possible mixed bag of others. In all of them the root notion is that of the necessitation of something by something else.

In speaking of necessity as a relation we should of course keep in mind what has been said about the existence of relations. In some cases, that of logical relations being a conspicuous example, we can simply deny their existence as anything to be added to the properties of the terms. In other cases, such as the causal, we can regard them as distinguishable elements in complex states of affairs. This does nothing however to affect the point that the terms are to be assumed to exist in the ordinary unqualified way. All that may need to be added is that there is a more complex state of affairs of one term necessitating the other.

The cases I have mentioned are, as I have said, the standard cases. They strongly suggest that the relational character of necessity is part of the concept. Talk of necessary existence might appear to be the exception. But if it is said of anything that it necessarily exists, we naturally take this as elliptical and ask what circumstances necessitate its existence. Put in terms of statements, we regard all statements of necessity as conditional, and rule out necessary existential ones.

There are cases admittedly where alleged necessary relations between existing items give us trouble. This is in fact the case as we have seen with the necessary dependence of ideas on minds and the mutual necessary dependence of substances and their properties. Difficulty in accepting such relations led to the idea that one or both of the items related had in themselves a dependent kind of existence, an idea we had to guard ourselves against by denying the dependence in one case, and denying in the other the dualism which gave rise to it. The trouble here, however, was with acceptance that there was a relation between existing items. There is no question that if we had accepted such a relation we would have assumed that this would have in no way implied the necessary existence of the items related, the minds or the ideas, the substances or the properties.

There is a line of argument we should mention which might be taken to counter this conclusion with the claim that there is at least something that necessarily exists. How, it may be said, can one item confer necessity on another unless it is somehow necessary in itself. It may of course have its necessity conferred on it by something else. But in the end there must be something which is necessary in itself, otherwise the whole sequence collapses. It is to this at least, it is said, that we have to attribute necessary existence. In the case of a causal sequence we speak of this as a first cause, but the principle applies whatever kind of necessitation we envisage. The position of such a first item of the series has often been attributed to God, whose existence has been taken to be a prime example of necessary existence.

The trouble with this line of argument is of course that it begs the question. It starts with the assumption that the character of the relation is somehow passed on to its consequent term, and must somehow be possessed by the antecedent term in order to be passed on. But this is something which has to be established, not assumed. Money no doubt circulates by being passed on from one person to another, but it is the money which is passed on, not its circulation. Similarly one thing after another is brought into existence through their necessary relationship but it is their existence which is 'passed on', not necessity. If a first cause exists and necessitates everything else, this will be because of the kind of entity it is, not because it has a reservoir of necessity to pour forth on the world.

When we speak of one thing necessitating another we should realise of course that the same relation can hold in reverse, the second thing necessitating the first. In the one case we speak of the second item,—call it B, as a necessary consequence of the first one,—call it A; in the other case we speak of A as a necessary condition of B, saying that B would not have existed without it. The distinction between necessary and sufficient conditions is familiar enough, the sufficient ones being those that necessitate consequences. Necessary conditions do not in themselves necessitate any consequences; they are rather necessitated by the existence of the consequences. But this reversal of the relationship doesn't add anything to the general idea of necessity as a relation, and therefore, we would suppose, it could give no further grounds for attributions of necessary existence.

There is however an argument, running to some extent parallel to the one we have dismissed, that suggests that this is not so. In this case the idea is that necessity is not so much passed on as passed back. Given some fact the existence of which no one doubts, we may be inclined to attribute necessary existence to a necessary condition of it. At the first step this may sound implausible. No one would suppose that oxygen exists necessarily because of the undoubted fact that I am alive. But if the fact is a very general fact, such as the existence of an ordered world, and we trace back the conditions to a first condition, the attribution of necessary existence to that first condition appears to have much more point. Those who attribute necessary existence to God are undoubtedly affected by the belief that the world as we know it would not have existed without such a being.

Such arguments which, following Kant, we have called transcendental arguments[2] certainly do have a point. They constitute an effective challenge to those who admit the fact but refuse to admit the existence of the condition. What we have to realise however is that it is the existence of the condition which is at stake, not its necessary existence. If it were true that the world would not exist without the existence of God, this would indeed be an argument for the existence of such a being—the familiar cosmological argument—but the necessity would be found not in God's existence but in God's relation to the world.

If we accept, then, that the root notion of necessity is that of the necessitation of one item by another, we have made a big step towards rejection of the view of necessary existence as a kind of existence. We have at least ruled out the attributing of necessary existence to any of the items related. The items exist and are of a certain nature. All that remains is the state of affairs of there being a certain relation of necessitation between them.

The question that now arises is whether we need to go any further in the characterising of necessity in order to defend the existence principle. For all that has been said so far necessity is still there as present in the world. There are in the world states of affairs of one thing necessitating another. Should we not ascribe necessary existence to those states of affairs?

The answer to this, I suggest, is that we have no reason to do so. If the items related do not necessarily exist, we cannot say that any relation between them does so and this applies even if that relation is one of necessitation. The state of affairs of a short circuit causally necessitating a fire might well exist, but no one would say it was necessary; it might not have happened. Given that humans do not live forever, Socrates being human logically necessitates his not living forever, but its doing so is not itself a necessary fact about the world, for after all Socrates might not have existed. If the relation of necessitation were a third item existing in addition to the terms related, we might indeed have to say that its existence was necessary. But relations, as I have argued, do not themselves exist, though they may (if external) be distinguished as elements in existing states of affairs. Necessity therefore could well remain as a relational element in the world without committing us to anything having a necessary existence.[3]

One way in which necessity has been said to be found in the world is through the possession by particular things of essential properties.

What is being spoken of here is not the possession by anything of some property or other, the necessity of which we discussed in the previous chapter. It is the necessary possession by some particular thing of a specific property, this being something which is obviously more controversial. Thus Socrates might be said to be only contingently snub-nosed but necessarily human. A table might be only contingently brown but necessarily made by some particular person from particular materials. If the things in question did not have these properties, it is said, they would not be the things they were.[4]

The issue whether there are such essential properties, bound up as it is with what constitutes a things identity, is not one we need go into here. All we need do is point out that, despite appearances, the necessity involved is still relational in character. It is being assumed that some particular thing exists, say Socrates, and that its particular property, say being human, also exists. What is being said is that the existence of the one necessitates the existence of the other. To call the property essential is to do no more than to say that it is so necessitated. From the point of view of the existence principle, therefore, if from no other, we can afford to live with essential properties.

At this point we might well claim that we need say no more about necessity. Merely pointing to its relational character precludes the idea of necessary existence. We are however left with it as a primitive concept of a certain relation holding between particular states of affairs. It looks as if we can only tell by inspection what necessitates what, what, for example, causes what, or what properties of things are essential rather than accidental. This being so, how are we to protect ourselves against the claim to find by inspection that in special cases, such as that of God, necessity is not relational at all and that we can attribute necessary existence to such a being? To buttress our argument, therefore, it would be as well to consider briefly what more can be said about necessity.

FURTHER ELUCIDATION OF NECESSITY

To begin with it should be pointed out that 'saying more about necessity' is a phrase which is intentionally vague. We should keep in mind that there is a spectrum ranging from what we might call elucidation to elimination, from drawing attention to certain further elements in the concept, to saying there is no such thing as necessity in the world at all. In between there is reduction which consists in pointing out necessity is a case of something which can be described in other terms.[5] In this spectrum there are no clear dividing lines and we should not be worried if what counts as necessity is to some extent a verbal matter.

In the case of causal necessity, one thing which some have done, in defiance of Hume, is to identify it with something that can be observed in particular cases. It may be said to be found, for example, in observed continuity or persistence of processes or in the experience of pressure on the surface of the body.[6] Some would say that such an appeal to observable facts would mean eliminating altogether any idea of necessity in causes, whereas others would regard it as a case of reduction. But the drawing of this dividing line is not our concern. What should rather concern us is that such proposals would apply only to causal necessity. If we are not going to treat the word 'necessity' as ambiguous as between its causal and other uses (and why should we?), this immediately rules out this account of necessity for anyone seeking a general elucidation of the concept.

As part of such a general elucidation a common move has been to say that necessity is not so much a relation between particular states of affairs as a relation between kinds of them, the relation between particulars being derivative from the relations between kinds. Necessity in short is to be regarded as accordance with law, different kinds of necessity corresponding to different kinds of law,—causal, logical, mathematical and so on. Taking this view makes it possible to introduce, as a special kind of law, rules established or conformed to by convention, an important case of these being the rules of language. Linguistic necessity can thus be added to causal, logical and mathematical necessity, and to any others such as those alleged to relate essential properties to kinds.[7]

The making of this move does not of course eliminate necessity from the world, it merely relocates it. Laws have to be distinguished from merely accidental generalisations and this is done by saying that the connections between universal properties or kinds is in their case a necessary one. We may however take comfort from the fact that there is nothing in this move to reintroduce the idea of necessary existence. Laws may be called necessary truths, but only in the sense that they assert certain relations of necessitation to hold between kinds of existents. They are universally quantified conditional statements, not existential ones. As such, they do not admittedly rule out the possibility that there are necessary existential statements but they do nothing to support it.

The relocation of necessity in laws does however open up certain possibilities of reduction which would effectively preclude any return to necessary existence. Of these there are notably two.

The first is the claim that all cases of necessity can be accounted for as matters of convention. The necessity of laws on this view can always be shown to be the necessity of rules laid down for human purposes, conspicuous among which are the rules of a language. Necessary relationships prescribed by these rules are said to be analytic. Necessity remains and is not eliminated, but is once more relocated, this time from the world we speak about to the way we speak about it.[8] This would give us secure protection against the idea that anything in the world necessarily existed.

No one doubts of course that there are such analytic necessities. Bachelors must be unmarried if we stick to that convention about the use of words. It is a further question however how far this way of accounting for necessity can be extended. It provides a plausible account of the essential properties of kinds, and there has been much support for the conventionality of mathematics. But as we move down through logical to causal necessities it becomes increasingly controversial. Which is a pity because, if accepted, it would be a simple way of settling the whole question of necessity once and for all.

The second way of precluding the possibility of necessary existence starts at the bottom end with causal necessity and has to be generalised upwards to other kinds short of the analytic. The necessity of causal laws, it has been said, does not consist in any intrinsic difference from other generalisations, but rather in their being parts of whole systems of laws which taken together constitute a theory. The denial of intrinsic difference is derived from Hume, the insistence on theoretical connection is what has been added. It is the theoretical connection which gives them explanatory power, and this, on such a view, is what we have in mind when we speak of their necessity.

If we take systems of logic and of mathematics as having to do with the very general structure of the world, it is not hard to see how such an account of necessity might be applied to them also. The difference is that being so general, they form parts of a great many theories, in the case of logic in fact of every theory. This gives them a special explanatory power, which leads to their being regarded as specially outstanding cases of necessity.

This neo-Humean way of accounting for necessity has, like the analytic way, met with many objections. They both however deserve mention as arguable ways of cutting short from the start any counter-claims about necessary existence which might be made so long as necessity is accepted as an unanalysable primitive concept. To accept it as such would not of course provide any argument in favour of there being necessary existence. If we are left with it, as primitive, we still have good reason for regarding it as a relational one only. So even if all further reductive accounts of necessity are rejected, appeal to it provides us with no positive reason for rejecting the existence principle.

POSSIBILITY

What I have said about necessity should apply in a parallel way to possibility since the two terms are interdefinable. What is possible is what is not necessarily not the case, and what is necessary is what is not possibly not the case. We might suppose therefore that we could dispose of the idea of possible existence in the same way as we have done for necessary existence by claiming that possibility is always relational in character. Just as necessity is to be found in one state of affairs necessitating another, so possibility consists in one state of affairs allowing another as possible. What is causally possible, we can say, is what is left open by the state of affairs which is the cause. What is logically possible is likewise what is left open by some existing state of the world. Whether an earthquake is possible depends on the state of the Earth's crust. Socrates' being mortal depends on his being human.

This may be all right as far as it goes, but it neglects an important difference between necessity and possibility when it comes to considerations about existence. Necessary existence, as we might put it, is something more than ordinary existence, possible existence something less. If we deny necessary existence of anything, we can fall back on admitting its existence. But if we deny possible existence we have nothing comparable to fall back on. To have a relation, we need existing terms, but the alleged relation of allowing-as-possible appears at first sight not to need for its second term anything with the ordinary kind of existence. Neither the possible earthquake nor the possible death of Socrates need be supposed to have occurred. Are we not then forced to retain possible existence as a kind of existence which these events have? After all, it may be said, there is something which is allowed as possible, but in general it is by no means something which actually exists.

The realm of what is allowed-as-possible does of course include the actual. Our trouble is that it extends beyond it. It has moreover to include possibilities of every kind. This means that its limits have to be taken as the limits of logical possibility, these being the widest limits.

There are many things which are logically possible which are not causally possible or even mathematically possible. If we think in terms of kinds of law, the laws of logic permit as possible anything which would fit into the most general structure of the world. Causal laws in contrast permit little as possible except the actual and if the universe is deterministic in character, the realm of the non-actual causally possible in fact shrinks to zero. So when the problem about this realm of possibilities is discussed, it is natural to present it in terms of what is logically possible.

I have stated the problem here as one about the possible existence of particular states of affairs. But it could be stated in terms of whole possible worlds and this is a way in which it is often presented. We may think of such worlds as numerically distinct both from our existing world and from each other. Our existing world, we should keep in mind, includes everything we normally speak of as existing, in the extent of space and time, as material or mental or whatever. And we take it to be the business of metaphysics to say something in general terms about this world. Other possible worlds will of course be in some respects similar to this, but unlike it in others, the remoteness of the possibility depending on the extent of the difference.

Whether we deal in terms of such possible worlds or in terms of particular possibilities, the same problem does of course remain about their ontological status, and it is this that concerns us. We have to ask how we can avoid assigning them some inferior kind of existence when they do not exist in our actual world and yet are needed as the second term of the allowing-as-possible relation.

Before we deal with this problem however there is still something more to be said about the idea of a possible world. We need to consider how talk of the possibility of such worlds fits with the view that possibility is always relational in character. We have to ask what allows a possible world as possible, seeing that it is conceived as a totality with nothing apart from it on which its possibility could depend. Are we not committed to taking its possibility as something which belongs to it simply in itself?

The answer to this is that in a way we are so committed but not in a way which conflicts with its possibility depending on circumstances. The point is that when we take a possible world as a totality, the circumstances on which the possibility depends have merely been incorporated into it. Some things in the world itself allow others to be within it and rule out still others as impossible. A tower's being square continues to rule out its being round and the Earth's crust being, say, forty miles thick rules out an earthquake. When we deal in possible worlds such contradictions and causal impossibilities become internal ones, but that is all.

Let us then return to the problem of how to account for the possible without allowing to it its own special kind of existence. As in previous cases we can point two ways in which this might be done. One is to say that what is possible simply exists without qualification like anything else. The other is to say that it does not exist at all. These are the alternatives we need to consider.

EXISTENCE OF THE POSSIBLE

The first of these alternatives undoubtedly has an air of paradox, but this of course is because we naturally want to distinguish sharply between what is the case and what only might be the case. That distinction clearly has to be preserved, but if this can be done, it may be argued that we can rid ourselves of any mystery about the realm of possibilities by raising its status to that of existing in the same way as what we call the actual world.

In its strongest form, this alternative takes us back to possible worlds. This time, however, talk of them ceases to be a mere device for representing the realm of possibilities, and becomes the assertion of their real existence. What is envisaged is that innumerably many of them exist, though only one of them is actual, this being the one we happen to inhabit.[9] If it is asked how they are to be distinguished from each other we can on this theory no longer use the standard criterion for distinguishing different worlds, that of their contents having different kinds of existence. Other ways of distinguishing them need to be found. Though on the old criterion they could be said to be mere parts of one great world of existing things which includes both the possible and the actual, they could nevertheless be distinguished, for example, by their being taken to exist in different times and spaces which have no relation to each other.[10]

Whatever may be said for or against this theory, it has the great virtue of dealing with the possible without breach of the existence principle. It might be thought that it fails to do this on the ground that it distinguishes the existent from the actual. Is this not virtually to allow a distinction between kinds of existence under another name? We might think of it as one of those hedging devices like that of distinguishing between the existent and the real or between existence and subsistence. But this is avoided if we accept the use of 'actual' as a token-reflexive word, expressing the attitude to any given world taken by its inhabitants. It is true that we normally identify the actual with the existent. But if we are prepared to modify our terminology, we need have no trouble on this score about the theory meeting our requirements.

The trouble that many will feel with this 'strong' theory is of course that it is ontologically extravagant. Not only does it assert to exist everything that is possible, including as it must everything that is logically possible. It also requires the existence of a whole new world for every difference in possibility, however minor. We may well ask whether there is any way of curtailing this extravagance while still retaining the idea of the existence of the possible.

There are various proposals which have allowed for this curtailment and we can present them as a series of weakening stages. The first stage is to retain possible worlds but allow them to overlap so that what exists in our world could also exist in others. The overlap might be small but it might also be very large, as in the case of a world differing from ours only by the amount of light emitted by a distant star. If sense could be made of such 'trans-world identity' as it has been called, it would avoid a great deal of duplication.[11]

The next stage would be to drop possible worlds altogether and revert to the existence of particular states of affairs allowed as possible by the state of the actual one. Taking the actual states of affairs as a base and considering only the existence of possibilities related to it would certainly narrow the field. But not too much should be made of the economy involved, since given the state of the actual world, the range of the logical possibilities alleged to exist would still be vast.

A third stage comes when the existence is denied of anything apart from the actual world but a niche is still found for possibilities within it. Thus they have been identified with various kinds of abstract objects such as sets of sentences, propositions or recombinations of basic existing items.[12] Such objects would have to be accepted as something which existed in the actual world, but at the same time firmly distinguished as possibilities.

We might consider here as a final stage a proposal that possibilities be fitted into the actual world as one kind of intentional object. Possible states of affairs are talked about, imagined, thought of, and so on. Why not, it may be said, treat them in the same way as images, fictions and the rest? Intentional objects, as we know, cause a problem for anyone concerned with existence. But the ways we have discussed of dealing with them could on this view be applied to envisaged possibilities also.[13]

The trouble with this last proposal is that it appears to go too far and to limit the range of possibility unduly. There are after all innumerable possibilities which have never been envisaged. While there are no imaginary objects which are not the objects of anyone's imagination, there are possible objects which no one has ever thought of. What is possible in fact is not what is thought of, but what it is possible to think of. So we are back with the possible existing somehow in independence of our attitudes, in whatever other way we decide to characterise it.

NON-EXISTENCE OF THE POSSIBLE

If we were to pursue the theme of economy there is no doubt that the best thing we could do would be to turn to the second way of avoiding treating possible existence as a distinct kind of existence. This is to deny that possible states of affairs exist at all, either in a separate unrelated realm or as parts of the actual world. On this view, the existent is to be identified with the actual and any reference to possibilities has to be accounted for in some other way.

If we take this alternative, it should be noted, we will not be able to say any longer that possibility, like necessity, is relational in character. Relations need terms, and what does not exist cannot be related to anything. But not too much should be made of this. We can still speak of what the existing circumstances allow as possible, and whatever sense we give to such statements, it must be such that the range of what is possible is determined by those circumstances.

The question is then what sense can be given to such statements. How can there be a range of possibilities with nothing in it? Clearly anyone who adopts this alternative is committed to a way of dealing with the non-existent which allows reference to it, but in a sense which does not imply that there is anything referred to. Such a way was mentioned in Ch. 4 in the discussion of non-existent objects.[14] It was the way taken by Bertrand Russell in his reply to Meinong when he said in effect that what was being done was to mention a kind and deny that it had instances. This idea can be applied to the case of the range of possibilities allowed by an actual situation. Thus given a round white tower there are many possibilities, including all logical possibilities, of shape, size, colour and so on, excluding only such states of affairs as its being both round and square or being both white and black. Apart from those instantiated in the tower, such as its being round and being white, these form a range of mere possibilities and of these it can be said quite generally that they do not exist. We thus avoid any commitment to objects occupying this range.

This account of the matter, it should be noted, fits very well with the idea that there are possibilities which have never been thought of and which therefore cannot be dealt with as intentional objects. We need have no idea what the permitted states of affairs might be or how many of them there might be if they did exist. All that we need have in mind is that given an actual state of affairs of a certain kind certain other kinds of states of affairs are left open as possible.

If this is what is required for making sense of the view that the possible does not exist, it must be admitted that it commits the holders of this view to regarding the possible as determined by law and only applicable derivatively to particular cases. If we stay on the level of particular states of affairs, we cannot just mention the range of the possibilities left open, say there is nothing in it and leave it at that. There is therefore a connection between adopting this alternative and accepting that it is part of the concept of possibility that it is lack of exclusion by law. Many admittedly would not regard this as a serious restriction, but it would be for those who would wish to combine this alternative with the view that we can distinguish without benefit of generalisation between the essential and the non-essential properties of particular things.

Here then in outline are the two alternative projects for dealing with possibilities within the limits of the dichotomy between their simply existing or not existing. Nothing has been added by either of them, we must now point out, about what this possibility consists in. The concept of possibility, that is to say, could well remain a primitive one, no reductive account of it being offered. If possible worlds exist they remain as possible and the assertion of their existence adds nothing to this.[15] If possibilities constitute a range with nothing in it, it remains a range of possibilities. Even where possibility is said to be determined by law, the laws remain as statements of necessary connections, that is, statements of the exclusion of the impossible. What we have been discussing here is not this issue but a distinct issue which arises only for the case of possibility, that about the existence of what is allowed as possible. What was said about the advantage of a further elucidation or reduction of the concept of necessity will of course apply *pari passu* to possibility as well. In each case it would be an advantage, though in each case it could be done without, without seriously affecting the concept of existence. In the case of possibilities, however, there is the extra problem about their existence and it is with that that we have been specifically concerned.

NOTES

1 In a book entitled *The Nature of Necessity* (Oxford, 1974), Alvin Plantinga says on p. 1 that on the question of what distinguished the necessary from the contingent, 'we must give example[s] and hope for the best'. He presents examples of logical necessity, 'broadly logical necessity', mathematical necessity and 'a host of homelier items' but excludes causal necessity as being necessity in another sense (he does not say why).

2 See Ch. 2, p. 24.

3 D.C. Williams, in his paper 'Necessary Facts' (*Review of Metaphysics*, 1962-3, pp. 601-625), claims there are necessary facts, because there are facts of necessary relation. While agreeing with his view about the conditional character of necessity, we need to point out that this way of expressing it is misleading.

4 For such a view, see Saul Kripke, *Naming and Necessity* (Oxford, 1980) p. 39 *et seq*. and p. 110 *et seq*. See also Plantinga, *The Nature of Necessity*, Chs 1-4, esp Ch. 4, Sect. 10. The necessity argued for here is sometimes called de re necessity, in contrast to de dicto necessity. We say no more about this disputed distinction, but the 'de re' tag clearly indicates a necessity belonging to the things themselves.

5 See Ch. 4, p. 71, on elimination and reduction.

6 For a source of such ideas, see A. Michotte, *The Perception of Causality* (Methuen, 1963). On continuity, see J. Mackie, *The Cement of the Universe* (Oxford, 1974) Ch. 8, esp. 218 *et seq*. On pressure see D. Armstrong, *A Materialist Theory of the Mind*, pp. 92 *et seq*. (Routledge, 1968).

7 On essential properties of kinds, as distinct from particulars see Kripke, *Naming and Necessity*, pp. 128-129, where the claim is considered that having a specific chemical composition is an essential property of water.

8 See Alan Sidelle, *Necessity, Essence and Individuation A Defense of Conventionalism* (Cornell, 1989) esp. Chs 4 and 5, for extended argument for this view.

9 David Lewis gives a systematic defence of this thesis in *On the Plurality of Worlds* (Blackwell, 1986). Note that on pp. 2-3 he explicitly rejects the idea that the possible and the actual differ in their 'manner of existing'. He has not the slightest idea, he says, what a difference in manner of existing is supposed to be.

10 This is the criterion used by David Lewis.

11 I by-pass here any general discussion of the issue about trans-world identity. On its problems, see e.g. Plantinga, *The Nature of Necessity*, Ch. 6; David Lewis, *On the Plurality of Worlds*, Sect. IV.

12 These are what David Lewis, *op. cit.*, Part 3, refers to and criticises as 'Ersatz' possible worlds, classifying them as linguistic, pictorial and magical. On recombination, see David Armstrong, *A Combinatorial Theory of Possibility* (Cambridge, 1989).

13 Nicholas Rescher, 'The Ontology of the Possible', in M. Munitz (ed.), *Logic and Ontology* (New York, 1973) is one who argues for possibilities existing as objects of thought.

14 See p. 69.

15 David Lewis claims to have avoided being left with modal primitives, and he argues against alternative views by saying they are left with them. It is hard to see how this claim fits with his rejection of impossible worlds. On this see W.G. Lycan's review of 'On the Plurality of Worlds', in *The Journal of Philosophy*, 1988, p. 46.

CHAPTER TEN

PERFECTION AND EXISTENCE

In considering the various ways in which the existence principle has been challenged, I have said nothing so far about the conspicuous challenge involved in the ontological argument for the existence of God. In the way it connects perfection with existence, this argument has involved such a conspicuous challenge to the principle that we need in conclusion to say something about it.

I mentioned it in Chapter 1 as a challenge to the fifth clause of the principle, the statement that existence is not a property. This was because the argument, on a standard account of it, is to the effect that a perfect being has to exist on pain of being less than perfect. This makes existence into a property of such a being, one, it is argued, that it has to have. It has various properties such as being omnipotent and omniscient, but if we do not add existence to them, it will be perfect no more. Therefore such a being exists.

A simple argument of this kind is to be found in the original statement by Anselm of Canterbury in the Twelfth Century.[1] It was developed clearly enough by Descartes in his Fifth Meditation.[2] And it was this argument, as most would agree, which was dealt with effectively by Kant when he pointed out that it was based on a misuse of the concept of existence.[3] In denying that existence was a 'real predicate', Kant was in fact rejecting the argument as in conflict with the existence principle. And if any further defence of this principle were necessary it could be based on the absurdity of abandoning it in this case. After all, if a perfect being did not exist it would have no properties at all, so it would not help if the additional property of existence were added to the list.

It might be supposed therefore that there is nothing more to be said about this argument, and that we should end by pointing to it as an interesting illustration of the troubles that arise when the existence principle is breached.

However, since the time of Anselm, the argument has generally been presented in a way which includes further elements not mentioned in the basic version I have given. We must ask how it fares when these elements are included and whether they give rise to versions of the argument which withstand the obvious criticism.

One of these elements is much to the fore in Anselm's original statement of the argument. He draws a distinction between existence in the understanding and existence in reality, and regards the argument as the making of a move from one to the other. God, as an absolutely perfect being, exists in the understanding in so far as we have a concept of such a being. The question is whether it also exists in reality. He argues that if it did not so exist, we would not have a concept of it, since our concept of it includes its existing. But we do have a concept of it. Hence, by *modus tollens*, it exists.

At first sight it looks as if this version of the argument merely moves us from one breach of the existence principle to another, from treating existence as a property to distinguishing kinds of existence. Once real existence is distinguished from existence in the understanding as one kind of existence only, treating it as a property becomes more plausible. But then we are left with the distinction between different kinds of existence. The argument becomes one from the one kind to the other, from intentional existence to real existence of a perfect being.

It should be pointed out straightaway that this version of the argument is in for trouble quite apart from its continuing to breach the existence principle. Even if we accept that there are these two kinds of existence, the making of this very distinction between kinds will undercut any possible argument from one to the other. Any relations between states of affairs which have different kinds of existence must remain, as I have maintained, completely mysterious. So we cannot infer from a perfect being having intentional existence that it has real existence, however much we choose to treat real existence as a property.[4]

We could of course avoid this particular trouble if we produced a version of the argument similar to this but without the idea that intentional objects which 'exist in the understanding' have any special kind of existence. We could give a representationalist account of our concept of a perfect being of the kind considered in Chapter 4. taking it to be some ordinarily existing item in the mind or the brain. Could we not then argue from the having of this concept to there being something corresponding to it in the world?

An argument of this kind would not conflict with the existence principle. It would in fact convert the ontological argument into a straightforward argument from effect to cause. It should be pointed out however that it remains a bad argument. There are no doubt cases where one can argue from representation to reality. We can do this if there is no other causal explanation of how we come by the representation. But in many cases there are other explanations, in particular that they are the result of mental construction out of other elements, as in the case of ghosts and chimaeras. Despite what has been claimed by Descartes and others,[5] there seems no reason to suppose that the concept of an infinite and perfect being has not been formed in this way by extrapolation and negation from concepts of the finite and the imperfect.

It seems then that to take the argument as one from existence in the understanding to existence in reality is not to improve on it. To give it substance we have to fall back, as Anselm fell back, on the original idea that perfection entails existence. The only difference, in this version, is that it is supposed to entail not just existence but real existence. And it is to be assumed that what is called real existence here is no different from what we ordinarily call existence.

So far nothing has been said about the necessity of the existence of a perfect and supreme being. Yet such necessity is often introduced in the exposition and refutation of the argument. Anselm considered himself to have proved that God *must* exist, that the non-existence of God was impossible. Descartes speaks of the necessity of the fact of God's existence. And Kant introduced his refutation by referring to the idea of an absolutely necessary being. This makes it look as if the conclusion of the argument involves us in yet another breach of the existence principle, this time the assumption of necessary existence as a special kind of existence.

It is of course possible to account for these references without supposing acceptance of this assumption. As far as these versions of the argument are concerned, all that need be meant is that the plain existence of a perfect being is logically necessitated by something else, its perfection or the thought of its perfection. It is a common practice to speak of a conclusion as necessary when we have shown that it is necessitated.

It has however been claimed that the absurdity of the basic version can be avoided by taking the argument to be one for the necessary existence of a perfect being in itself and not just for its existence being necessitated.[6] Let us agree, it may be said, that existence is not a property, but this need not apply to necessary existence, any more than it needed to apply to real existence on the previous version. In this version the argument relies no longer on the superiority of the existent over the non-existent. It relies instead on the superiority of the necessarily existent over the contingently existent. Perfection requires necessary existence; anything the non-existence of which is possible would not suffice. Hence a perfect being necessarily exists.

This version of the argument brings us back to the question about necessary existence considered in the previous chapter. In this version the argument would collapse if necessary existence were not taken at its face value as a kind of existence. Yet I have claimed that all alleged cases of necessary existence can be accounted for in terms of necessitation by something else. In particular I have claimed that the alleged necessary existence of God as a first cause can be accounted for in terms of the claim that God is a necessary condition of everything else. But perhaps here we have come at last across a reason for thinking that in this one case at least we have a genuine instance of necessary existence which cannot be accounted for in the same way? If so, we have found at last a direct challenge to the existence principle in its third clause if not its fifth.

The natural thing to do here is merely to repeat the claim that what is meant by a perfect being necessarily existing is that its existence is necessitated by its perfection. If this is denied, as it has been by proponents of this version of the argument,[7] all that can be done is to ask these proponents to say something more if they can about what the possession of necessary existence involves. To say it involves that the non-existence of such a being is impossible, or that it exists in all possible worlds, is not of course to take the matter further. So what more can be said?

One thing which has been said is that nothing less than the necessary existence of a supreme being would satisfy those who take up a religious attitude. A God, it is said, who 'just happens' to exist would not be the object of religious reverence or worship. For this there has to be a being whose presence is not a mere fact, but is unavoidable. With this is associated the requirement that it be absolutely independent, unlimited and eternal, and therefore that there is nothing which could prevent its existence.[8]

The question here is what is meant by something just happening to exist. It is intended without doubt to rule out necessary existence, but what more does it imply? One thing it might mean is that the existence of the thing in question is not necessitated by anything else. But this fits exactly with its being independent and unlimited. Wherever there are dependent things, there must in the last resort be something which is independent. It could be the first item in a series, or according to some ways of thinking it could be the whole of which everything else is a part. But whichever it is it cannot be explained further in terms of other things, so we just have to accept that it exists and leave it at that. The question is why this should not apply to God, the unconditioned and eternal being which is the object of religious reverence, without necessity being brought into it. To admit this would of course be to admit that the non-existence of such a being is possible. But there seems no reason why an attitude of religious reverence should require that the agnostic and the atheist should be charged with inconsistency.

The objection to saying that God just happens to exist is no doubt made more plausible by a further implication. This is the suggestion it contains that God's existence is a matter of chance, or merely accidental. This implies that there is already some necessarily connected scheme of things in the world, but it is one into which the existence of a perfect and independent being does not fit. To be a matter of chance is to be an item without causal explanation in an indeterministic world, like the breakdown of a radioactive particle. To be an accident is to be a point of intersection of two or more relatively independent causal chains. To say that God exists without attributing any necessity to that existence, is not to commit oneself to any of those things. While such a being is not regarded as being itself necessitated, it is nevertheless regarded as a necessary condition for the existence of all other things, and therefore the central point in a system of necessitation. The status of chance events depends admittedly on one's views about the ultimacy of statistical laws. But no issue of this kind would be relevant for consideration about the existence of God.

It seems then that neither the appeal to the religious attitude nor the appeal to the independent and unconditioned character of God take us any further in the understanding of necessary existence. In this predicament it is of course always possible to say that 'necessity' has a different meaning in different contexts, and that therefore reference to necessary existence has a point in the religious context. As it has been put, following Wittgenstein, this language-game is played, at least in the Jewish and Christian traditions.[9] However, which language-games are played in which traditions is not relevant for those concerned with the implications of the basic concept of existence. Their concern, when it comes to necessity, is to identify a similarly basic concept, and to see how it fits with that of existence. The conclusion is that it fits only if necessary existence is the same as necessitated existence.

Once necessary existence ceases to be regarded as a special kind of existence, the third version of the argument no longer holds. We are back again to the basic version which depends on existence plain and simple being regarded as a property. And about this version, as I have suggested, there is nothing more to be said other than that it is an excellent illustration of the dire consequences which arise when one ignores the existence principle.

I have concluded in this chapter with a conspicuous case of a departure from the principle, one which has clearly unacceptable consequences and which can be avoided merely by rejecting the argument.

What has emerged in the course of the preceding chapters is that there are many less conspicuous cases. My intention has not been to systematise them or to attempt a complete list, but rather to present them as important cases in which the requirements of the concept of existence have been ignored. My object has been to bring them out into the open as unacceptable ways of evading certain metaphysical problems, and where necessary to indicate ways in which the problems they evade can be dealt with within the limits set by the existence principle.

NOTES

1 Anselm, *Proslogion*, Chs II-IV translated by S.N. Deane in *Anselm's Basic Writings* (2nd ed., 1962, La Salle, Illinois).

2 Descartes, *Meditations on First Philosophy*, translated by Anscombe and Geach in *Descartes, Philosophical Writings* (Nelson, 1954) pp. 103-105.

3 Kant, *Critique of Pure Reason*, M.P. B620-B630.

4 See W.P. Alston, 'The Ontological Argument Revisited', in *The Philosophical Review*, Vol. LXIX, 1960, pp. 452-474, for the presentation of the argument in this form and its rejection on this ground. Alston is one who is happy to accept the view that there are different 'modes of existence' but points to the problem it raises for the argument.

5 See Descartes, Meditation II,. *op. cit.*, pp. 83-87.

6 See Norman Malcolm, 'Anselm's Ontological Argument', in *The Philosophical Review*, Vol. LXIX (1960) pp. 41-62 for presentation and defence of this version of the argument.

7 See Norman Malcolm, *op. cit.*, Sect. II (p. 50) 'we are not to think that "God necessarily exists" means that it follows necessarily from something that God exists contingently'.

8 J.N. Findlay, 'Can God's Existence be Disproved?', in *Mind* (1948), points to these requirements, and taking all necessity to be analytic, regards them as showing that such a being does not exist. Norman Malcolm, *op. cit.*, uses them, on the other hand, in support of a necessary existence, the one he claims can be established by an ontological argument.

9 See Norman Malcolm, *op. cit.*, p. 56.

REFERENCES

A LIST OF WORKS MENTIONED IN THE TEXT

Alexander, S. (1920), *Space, Time and Deity*. London.

Alston, W.P. (1960), 'The Ontological Argument Revisited'. In *The Philosophical Review*, **69**.

Anderson, J. (1962), *Studies in Empirical Philosophy*. Sydney.

Annas, J. (1981), *An Introduction to Plato's Republic*. Oxford.

Anscombe, G.E.M. (1987-8), 'Existence and Truth'. In *Aristotelian Society Proceedings*, NS **LXXXVIII**. London.

Anselm (1962), *Proslogion*. Tr. Deane. In *Anselm's Basic Writings*, 2nd ed. La Salle.

Aristotle (1937), *On Interpretation*. In W.D. Ross (ed.), *The Works of Aristotle*. Oxford.

Armstrong, D. (1961), *Perception and the Physical World*. London.

——— (1968), *A Materialist Theory of the Mind*. London.

——— (1989), *Universals*. Westview.

——— (1989), *A Combinatorial Theory of Possibility*. Cambridge.

Augustine (1948), *Confessions*. In Whitney J. Oates (ed.), *Basic Writings of Saint Augustine*. New York.

Austin, J. (1962), *Sense and Sensibilia*. Oxford.

Ayer, A.J. (1946), *Language, Truth and Logic*, 2nd ed. London.

Bergson, H. (1911), *Creative Evolution*. London.

Berkeley, G. (1965), *The Principles of Human Knowledge*. In *Berkeley's Philosophical Writings*. New York.

Bradley, F.H. (1893), *Appearance and Reality*. London.

Brentano, F. (1955), *Psychologie vom Empirischen Standpunkt*, Vol. I, 1874. Reprinted by Felix Meiner. Hamburg.

Campbell, K. (1990), *Abstract Particulars*. Oxford.

Cheyne, C. (1991), 'Reduction, Elimination and Firewalking'. In *Philosophy of Science*, **58**.

Chisholm, R.M. (ed.) (1960), *Realism and the Background of Phenomenology*. Illinois.

——— (1972), 'Beyond Being and Non-Being'. In *New Readings in Philosophical Analysis*. New York.

——— (1976), 'Homeless Objects'. In *Revue Internationale de Philosophie*, **27**.

Churchland, P. (1981), 'Eliminative Materialism and the Propositional Attitudes'. In *Journal of Philosophy*, **79**.

——— (1984), *Matter and Consciousness*. Cambridge, Mass.

Cross, R.C. and A.D. Woozley (1964), *Plato's Republic*. London.

Dennett, D. (1969), *Content and Consciousness*. London.

Descartes, R. (1954), *Meditations on First Philosophy*. In *Descartes' Philosophical Writings*, trans. Anscombe and Geach. London.

Devitt, M. (1984), *Realism and Truth*. Princeton.

Dummett, M. (1975), 'Wang's Paradox'. In *Synthese*, **30**.

——— (1978), *Truth and Other Enigmas*. London.

Feigl, Sellars and Lehrer (eds) (1962), *New Readings in Philosophical Analysis*. New York.

Findley, J.N. (1941), 'Time: A Treatment of Some Puzzles'. In *Australasian Journal of Psychology and Philosophy*, **19**.

——— (1948), 'Can God's Existence be Disproved?'. In *Mind*, **57**.

Fine, A. (1975), 'Vagueness, Truth and Logic'. In *Synthese*, **30**.

Fodor, J. (1981), *Representations*. Brighton.

Forrest, P. (1988), *Quantum Metaphysics*. Oxford.

Gale, R.M. (ed.) (1969), *The Philosophy of Time*. Anchor Books.

Haack, S. (1974), *Deviant Logic*. Cambridge.

Heidegger, M. (1953), *Sein und Zeit*. Tübingen.

Heller, M. (1988), 'Vagueness and the Standard Ontology'. In *Nous*, **22**.

Hollis, M. and S. Lukes (eds) (1982), *Rationality and Relativism*. Oxford.

Hume, D. (1978), *Treatise of Human Nature*, ed. L.A. Selby-Bigge, 2nd ed. revised by P.H. Nidditch. Oxford.

Husserl, E. (1982), *Ideas*, Bk. I, trans. F. Kersten. The Hague.

——— (1870), *Logical Investigations*, trans. J.N. Findlay. London.

Jackson, F. (1977), *Perception*. Cambridge.

Kant, I. (1963), *Critique of Pure Reason*, trans. N. Kemp-Smith. London.

Kim, J. (1974), 'Noncausal Connections'. In *Nous*, **8**.

——— (1976), 'Events as Property Exemplications'. In M. Brand and D. Walton (eds), *Action Theory*. Dordrecht.

Körner, S. (1970), *Categorial Frameworks*. Oxford.

Kripke, S. (1980), *Naming and Necessity*. Oxford.

Krips, H. (1987), *The Metaphysics of Quantum Theory*. Oxford.

Kuhn, T. (1970), *The Structure of Scientific Revolutions*, 2nd ed. Chicago.

Lenin, V.I. (1928), *Materialism and Empirio-Criticism*. London.

Lewis, D. (1986), *On the Plurality of Worlds*. Oxford.

Lloyd, G. (1978), 'Time and Existence'. In *Philosophy*, **53**.

Locke, J. (1975), *Essay Concerning Human Understanding*, ed. P.H. Nidditch. Oxford.

Lukes, S. (1982), 'Relativism in its Place'. In M. Hollis and S. Lukes (eds), *Rationality and Relativism*. Oxford.
Lycan, W.G. (1988), Review of D. Lewis 'On the Plurality of Worlds'. In *Journal of Philosophy*, **85**.
Mackie, J. (1976), 'The Riddle of Existence'. In *Aristotelian Society Supplementary*, **L**.
——— (1974), *The Cement of the Universe*. Oxford.
Malcolm, N. (1960), 'Anselm's Ontological Argument'. In *Philosophical Review*, **69**.
McTaggart, J.M.E. (1927), *The Nature of Existence*. Cambridge.
Mannheim, K. (1936), *Ideology and Utopia*. London.
Martin, C.B. (1980), 'Substance Substantiated'. In *Australasian Journal of Philosophy*, **58**.
Meinong, M. (1960), *The Theory of Objects*, reprinted in R.M. Chisholm, *Realism and the Background of Phenomenology*. Illinois.
Mellor, D.H. (1981), *Real Time*. Cambridge.
Michotte, A. (1963), *The Perception of Causality*. London.
Miller, B. (1986), '"Exists" and Existence'. In *Review of Metaphysics*, **XL**.
Moore, G.E. (1922), *Philosophical Studies*. London.
Moreland, J.P. (1989), 'Keith Campbell and the Trope View of Predication'. In *Australasian Journal of Philosophy*, **67**.
Nagel, T. (1986), *The View from Nowhere*. Oxford.
Parsons, T. (1980), *Non-Existent Objects*. Boston.
Passmore, J. (1961), *Philosophical Reasoning*. London.
Plantinga, A. (1974), *The Nature of Necessity*. Oxford.
Plato (1955), *The Republic*, trans. H.D.P. Lee. Penguin.
Popper, K.R. (1972), *Objective Knowledge*. Oxford.
Priest, G. (1987), *In Contradiction*. The Hague.
Prior, A.N. (1968), *Time and Tense*. Oxford.
Putnam, H. (1975), *The Logic of Quantum Mechanics*. In *Philosophical Papers*, Vol. I. Cambridge.
Quine, W. van O. (1961), *Two Dogmas of Empiricism* (*Philosophical Review*, **60** [1951]) as reprinted in *From a Logical Point of View*, 2nd ed. Harvard.
Quinton, A. (1957-8), 'Properties and Classes'. In *Proceedings of the Aristotelian Society*, NS **LVIII**.
Rescher, N. (1969), *Many Valued Logic*. New York.
——— (1972), 'The Ontology of the Possible'. In M. Munitz (ed.), *Logic and Ontology*. New York.
——— (1978), *Studies in Ontology*. Oxford.
Romney, G. (1977-8), 'Temporal Points of View'. In *Proceedings of the Aristotelian Society*, NS **LXXVIII**.

Routley, R. & V. (1973), 'Rehabilitating Meinong's Theory of Objects'. In *Revue Internationale de Philosophie*, **27**.
Russell, B. (1904), 'Meinong's Theory of Complexes and Assumptions'. In *Mind*, **13**.
———— (1905), 'On Denoting'. In *Mind*, **14**.
———— (1923), 'Vagueness'. In *Australasian Journal of Psychology and Philosophy*, **1**.
Schräder, G. (1949), 'The Thing-in-Itself in Kantian Philosophy'. In *Review of Metaphysics*, **3**.
Schwartz, J. (1991), 'Reduction, Elimination and the Mental'. In *Philosophy of Science*, **58**.
Sellars, W. (1962), *Time and World Order*. In *Minnesota Studies in the Philosophy of Science*, III. Minneapolis.
Sidelle, A. (1989), *Necessity, Essence and Individuation*. Cornell.
Simons, P. (1987), *Parts: A Study in Ontology*. Oxford.
Sober, E. (1982), 'Realism and Independence'. In *Nous*, **16**.
Strawson, P.F. (1950), 'On Referring'. In *Mind*, **59**.
———— (1964), 'Identifying Reference and Truth-Values'. In *Theoria*, **30**.
Taylor, B. (1985), *Modes of Occurrence*. Oxford.
Tennant, N. (1987), *Anti-Realism and Logic*. Oxford.
Unger, P. (1979), 'There are no Ordinary Things'. In *Synthese*, **44**.
Vlastos, G. (1973), *Platonic Studies*. Princeton.
Wheeler, S.E. (1979), 'On That which is Not'. In *Synthese*, **44**.
Williams, C.J.F. (1981), *What is Existence?* Oxford.
Williams, D.C. (1962-3), "Necessary Facts'. In *Review of Metaphysics*, **16**.
———— (1966), *Principles of Empirical Realism*. Illinois.
Williamson, T. (1987-8), 'Equivocation and Existence'. In *Proceedings of the Aristotelian Society*, NS **LXXXVIII**.
Winch, P. (1964), 'Understanding a Primitive Society'. In *American Philosophical Quarterly*, **1**.
Wittgenstein, L. (1922), *Tractatus Logico-Philosophicus*. London.
———— (1968), *Philosophical Investigations*. Oxford.
Wolterstorff, N. (1961), 'Referring and Existing'. In *Philosophical Quarterly*, **11**.

INDEX

PHILOSOPHICAL STUDIES SERIES

1. Jay F. Rosenberg: *Linguistic Representation.* 1974 ISBN 90-277-0533-X
2. Wilfrid Sellars: *Essays in Philosophy and Its History.* 1974 ISBN 90-277-0526-7
3. Dickinson S. Miller: *Philosophical Analysis and Human Welfare.* Selected Essays and Chapters from Six Decades. Edited with an Introduction by Lloyd D. Easton. 1975 ISBN 90-277-0566-6
4. Keith Lehrer (ed.): *Analysis and Metaphysics.* Essays in Honor of R. M Chisholm. 1975 ISBN 90-277-0571-2
5. Carl Ginet: *Knowledge, Perception, and Memory.* 1975 ISBN 90-277-0574-7
6. Peter H. Hare and Edward H. Madden: *Causing, Perceiving and Believing.* An Examination of the Philosophy of C. J. Ducasse. 1975 ISBN 90-277-0563-1
7. Hector-Neri Castañeda: *Thinking and Doing.* The Philosophical Foundations of Institutions. 1975 ISBN 90-277-0610-7
8. John L. Pollock: *Subjunctive Reasoning.* 1976 ISBN 90-277-0701-4
9. Bruce Aune: *Reason and Action.* 1977 ISBN 90-277-0805-3
10. George Schlesinger: *Religion and Scientific Method.* 1977 ISBN 90-277-0815-0
11. Yirmiahu Yovel (ed.): *Philosophy of History and Action.* Papers presented at the First Jerusalem Philosophical Encounter (December 1974). 1978 ISBN 90-277-0890-8
12. Joseph C. Pitt (ed.): *The Philosophy of Wilfrid Sellars: Queries and Extensions.* 1978 ISBN 90-277-0903-3
13. Alvin I. Goldman and Jaegwon Kim (eds.): *Values and Morals.* Essays in Honor of William Frankena, Charles Stevenson, and Richard Brandt. 1978 ISBN 90-277-0914-9
14. Michael J. Loux: *Substance and Attribute.* A Study in Ontology. 1978 ISBN 90-277-0926-2
15. Ernest Sosa (ed.): *The Philosophy of Nicholas Rescher.* Discussion and Replies. 1979 ISBN 90-277-0962-9
16. Jeffrie G. Murphy: *Retribution, Justice, and Therapy.* Essays in the Philosophy of Law. 1979 ISBN 90-277-0998-X
17. George S. Pappas (ed.): *Justification and Knowledge.* New Studies in Epistemology. 1979 ISBN 90-277-1023-6
18. James W. Cornman: *Skepticism, Justification, and Explanation.* With a Bibliographic Essay by Walter N. Gregory. 1980 ISBN 90-277-1041-4
19. Peter van Inwagen (ed.): *Time and Cause.* Essays presented to Richard Taylor. 1980 ISBN 90-277-1048-1
20. Donald Nute: *Topics in Conditional Logic.* 1980 ISBN 90-277-1049-X
21. Risto Hilpinen (ed.): *Rationality in Science.* Studies in the Foundations of Science and Ethics. 1980 ISBN 90-277-1112-7
22. Georges Dicker: *Perceptual Knowledge.* An Analytical and Historical Study. 1980 ISBN 90-277-1130-5
23. Jay F. Rosenberg: *One World and Our Knowledge of It.* The Problematic of Realism in Post-Kantian Perspective. 1980 ISBN 90-277-1136-4
24. Keith Lehrer and Carl Wagner: *Rational Consensus in Science and Society.* A Philosophical and Mathematical Study. 1981 ISBN 90-277-1306-5
25. David O'Connor: *The Metaphysics of G. E. Moore.* 1982 ISBN 90-277-1352-9
26. John D. Hodson: *The Ethics of Legal Coercion.* 1983 ISBN 90-277-1494-0
27. Robert J. Richman: *God, Free Will, and Morality.* Prolegomena to a Theory of Practical Reasoning. 1983 ISBN 90-277-1548-3

PHILOSOPHICAL STUDIES SERIES

28. Terence Penelhum: *God and Skepticism.* A Study in Skepticism and Fideism. 1983 ISBN 90-277-1550-5
29. James Bogen and James E. McGuire (eds.): *How Things Are.* Studies in Predication and the History of Philosophy of Science. 1985 ISBN 90-277-1583-1
30. Clement Dore: *Theism.* 1984 ISBN 90-277-1683-8
31. Thomas L. Carson: *The Status of Morality.* 1984 ISBN 90-277-1619-9
32. Michael J. White: *Agency and Integrality.* Philosophical Themes in the Ancient Discussions of Determinism and Responsibility. 1985 ISBN 90-277-1968-3
33. Donald F. Gustafson: *Intention and Agency.* 1986 ISBN 90-277-2009-6
34. Paul K. Moser: *Empirical Justification.* 1985 ISBN 90-277-2041-X
35. Fred Feldman: *Doing the Best We Can.* An Essay in Informal Deontic Logic. 1986 ISBN 90-277-2164-5
36. G. W. Fitch: *Naming and Believing.* 1987 ISBN 90-277-2349-4
37. Terry Penner: *The Ascent from Nominalism.* Some Existence Arguments in Plato's Middle Dialogues. 1987 ISBN 90-277-2427-X
38. Robert G. Meyers: *The Likelihood of Knowledge.* 1988 ISBN 90-277-2671-X
39. David F. Austin (ed.): *Philosophical Analysis.* A Defense by Example. 1988 ISBN 90-277-2674-4
40. Stuart Silvers (ed.): *Rerepresentation.* Essays in the Philosophy of Mental Representation. 1988 ISBN 0-7923-0045-9
41. Michael P. Levine: *Hume and the Problem of Miracles.* A Solution. 1989 ISBN 0-7923-0043-2
42. Melvin Dalgarno and Eric Matthews (eds.): *The Philosophy of Thomas Reid.* 1989 ISBN 0-7923-0190-0
43. Kenneth R. Westphal: *Hegel's Epistemological Realism.* A Study of the Aim and Method of Hegel's *Phenomenology of Spirit.* 1989 ISBN 0-7923-0193-5
44. John W. Bender (ed.): *The Current State of the Coherence Theory.* Critical Essays on the Epistemic Theories of Keith Lehrer and Laurence BonJour, with Replies. 1989 ISBN 0-7923-0220-6
45. Roger D. Gallie: *Thomas Reid and 'The Way of Ideas'.* 1989 ISBN 0-7923-0390-3
46. J-C. Smith (ed.): *Historical Foundations of Cognitive Science.* 1990 ISBN 0-7923-0451-9
47. John Heil (ed.): *Cause, Mind, and Reality.* Essays Honoring C. B. Martin. 1989 ISBN 0-7923-0462-4
48. Michael D. Roth and Glenn Ross (eds.): *Doubting.* Contemporary Perspectives on Skepticism. 1990 ISBN 0-7923-0576-0
49. Rod Bertolet: *What is Said.* A Theory of Indirect Speech Reports. 1990 ISBN 0-7923-0792-5
50. Bruce Russell (ed.): *Freedom, Rights and Pornography.* A Collection of Papers by Fred R. Berger. 1991 ISBN 0-7923-1034-9
51. Kevin Mulligan (ed.): *Language, Truth and Ontology.* 1992 ISBN 0-7923-1509-X
52. Jesús Ezquerro and Jesús M. Larrazabal (eds.): *Cognition, Semantics and Philosophy.* Proceedings of the First International Colloquium on Cognitive Science. 1992 ISBN 0-7923-1538-3
53. O.H. Green: *The Emotions.* A Philosophical Theory. 1992 ISBN 0-7923-1549-9
54. Jeffrie G. Murphy: *Retribution Reconsidered.* More Essays in the Philosophy of Law. 1992 ISBN 0-7923-1815-3

PHILOSOPHICAL STUDIES SERIES

55. Phillip Montague: *In the Interests of Others*. An Essay in Moral Philosophy. 1992 ISBN 0-7923-1856-0
56. Jacques-Paul Dubucs (ed.): *Philosophy of Probability*. 1993 ISBN 0-7923-2385-8
57. Gary S. Rosenkrantz: *Haecceity*. An Ontological Essay. 1993 ISBN 0-7923-2438-2
58. Charles Landesman: *The Eye and the Mind*. Reflections on Perception and the Problem of Knowledge. 1994 ISBN 0-7923-2586-9
59. Paul Weingartner (ed.): *Scientific and Religious Belief*. 1994 ISBN 0-7923-2595-8
60. Michaelis Michael and John O'Leary-Hawthorne (eds.): *Philosophy in Mind*. The Place of Philosophy in the Study of Mind. 1994 ISBN 0-7923-3143-5
61. William H. Shaw: *Moore on Right and Wrong*. The Normative Ethics of G.E. Moore. 1995 ISBN 0-7923-3223-7
62. T.A. Blackson: *Inquiry, Forms, and Substances*. A Study in Plato's Metaphysics and Epistemology. 1995 ISBN 0-7923-3275-X
63. Debra Nails: *Agora, Academy, and the Conduct of Philosophy*. 1995 ISBN 0-7923-3543-0
64. Warren Shibles: *Emotion in Aesthetics*. 1995 ISBN 0-7923-3618-6
65. John Biro and Petr Kotatko (eds.): *Frege: Sense and Reference One Hundred Years Later*. 1995 ISBN 0-7923-3795-6
66. Mary Gore Forrester: *Persons, Animals, and Fetuses*. An Essay in Practical Ethics. 1996 ISBN 0-7923-3918-5
67. K. Lehrer, B.J. Lum, B.A. Slichta and N.D. Smith (eds.): *Knowledge, Teaching and Wisdom*. 1996 ISBN 0-7923-3980-0
68. Herbert Granger: *Aristotle's Idea of the Soul*. 1996 ISBN 0-7923-4033-7
69. Andy Clark, Jesús Ezquerro and Jesús M. Larrazabal (eds.): *Philosophy and Cognitive Science: Categories, Consciousness, and Reasoning*. Proceedings of the Second International Colloquium on Cogitive Science. 1996 ISBN 0-7923-4068-X
70. J. Mendola: *Human Thought*. 1997 ISBN 0-7923-4401-4
71. J. Wright: *Realism and Explanatory Priority*. 1997 ISBN 0-7923-4484-7
72. X. Arrazola, K. Korta and F.J. Pelletier (eds.): *Discourse, Interaction and Communication*. Proceedings of the Fourth International Colloquium on Cognitive Science. 1998 ISBN 0-7923-4952-0
73. E. Morscher, O. Neumaier and P. Simons (eds.): *Applied Ethics in a Troubled World*. 1998 ISBN 0-7923-4965-2
74. R.O. Savage: *Real Alternatives, Leibniz's Metaphysics of Choice*. 1998 ISBN 0-7923-5057-X
75. Q. Gibson: *The Existence Principle*. 1998 ISBN 0-7923-5188-6

KLUWER ACADEMIC PUBLISHERS – DORDRECHT / BOSTON / LONDON

GPSR Compliance
The European Union's (EU) General Product Safety Regulation (GPSR) is a set of rules that requires consumer products to be safe and our obligations to ensure this.

If you have any concerns about our products, you can contact us on

ProductSafety@springernature.com

In case Publisher is established outside the EU, the EU authorized representative is:

Springer Nature Customer Service Center GmbH
Europaplatz 3
69115 Heidelberg, Germany

www.ingramcontent.com/pod-product-compliance
Lightning Source LLC
LaVergne TN
LVHW011802200926
848484LV00003B/762
9780792351887